Fundamentals of Quantum and Classical Optics

Fundamentals of Quantum and Classical Optics:

Squeezed State and Super Resolutions

By

Yacob Ben-Aryeh

Cambridge Scholars Publishing

Fundamentals of Quantum and Classical Optics: Squeezed State and Super Resolutions

By Yacob Ben-Aryeh

This book first published 2026

Cambridge Scholars Publishing

Lady Stephenson Library, Newcastle upon Tyne, NE6 2PA, UK

British Library Cataloguing in Publication Data
A catalogue record for this book is available from the British Library

ISBN: 978-1-0364-6804-0
ISBN (Ebook): 978-1-0364-6805-7

TABLE OF CONTENTS

LIST OF ILLUSTRATIONS

ACKNOWLEDGMENTS

I wish to express my deep gratitude to my colleagues and collaborators in the Department of Physics at the Technion – Israel Institute of Technology and in other institutions. Their stimulating discussions and critical feedback have enriched this work on quantum optics. Special topics in classical optics are studied with the insight of the present author.

I am especially indebted to Professor Adi Mann, whose scientific rigor and generous guidance were instrumental throughout the research of this book.

I would also like to thank my editor, Dr. Noa Burshtein, for her professionalism, insightful suggestions, and dedication to bringing this project to fruition.

Much of the research presented here was supported by the excellent facilities and intellectual environment provided by the Technion, and I am grateful for the opportunities this institution afforded me. The present study was supported by Technion-Mosad under grant No. 2007156.

Finally, I wish to express my heartfelt thanks to my daughter, Nirit Ben-Aryeh Debby, whose love, patience, and encouragement sustained me through the long process of writing and revision. Her curiosity and spirit continue to inspire me every day.

Conflict of Interest

The author declares no conflict of interest. Artificial AI-assisted technology was not used.

INTRODUCTION

The aim of the present book is to present central topics in quantum and classical optics, and to develop the theoretical framework for each topic from first principles familiar to graduate students and researchers. The analysis is simplified as much as possible without losing the important physical features of each system. For cases in which the analysis becomes too complex, we refer the reader to the relevant works of the present author and others. While the extensive body of literature on these topics is not fully covered, a focused selection of references is provided to support and complement the mathematical derivations presented here.

The analysis in this book falls into two main categories. The first focuses on the fundamentals of quantum optics. Although the material introduced in these sections is inherently complex, we adopt the most straightforward approach for their interpretations. In some of these topics, where conceptual issues arise, the author presents a specific perspective to address these issues. The second category includes the examination of selected topics from different scientific areas, which are typically absent from textbooks on quantum and classical optics, and may be of particular interest to researchers and graduate students.

In the second part of the book, five complex quantum optics systems are treated. In Chapter 10, we explore how squeezed light interacting with atoms can produce squeezed atomic states and examine the implications of such interactions. In Chapter 11, it is shown that local dipole-dipole interactions lead to changes in the Maxwell-Bloch equations of dense two-level systems, resulting in intrinsic bistability effects. In Chapter 12, the separability and entanglement of 2, 3, and 4 qubits are treated, a core topic in the field of quantum computation. In Chapter 13, we treat the use of pseudo-Hermitian Hamiltonians for physical systems exhibiting parity-time symmetry and possessing real eigenvalues. In Chapter 14, we treat quantum engines, where unitary transformations of their mixed states can lead to geometric phases.

In the third part of the book, we show how super resolutions can be obtained through label-free nanoscale object scattering (i.e., without fluorescent atoms, which can enhance resolutions, but can distort the imaging process). In Chapter 15, we treat super-resolution effects observed in hot spots, where extremely high electromagnetic intensity is confined to a small region, enabling Raman spectroscopy at the single molecule level. In Chapter 16, we examine super-resolution obtained by microsphere systems, where their explicit geometry helps the analysis.

We hope that the additional topics in this book will raise scientific interest beyond the interest in the fundamentals of quantum optics.

List of Abbreviations

AB	Aharonov and Bohm
AN	Anti-normally
BCH	Baker-Campbell-Hausdorf
BS	Beam splitter
CR	Commutation relations
CS	Coherent state
DDA	Discrete dipole approximation
EF	Enhancement factor
EM	Electromagnetic
EPR	Einstein-Podolsky-Rosen
EW	Entanglement witnesses
FS	Fully separable
HS	Hilbert–Schmidt
IOB	Intrinsic optical bistability
IS	Intelligent states
JCM	Jaynes–Cummings model
LHV	Local hidden variable
LO	Local oscillator
NA	Numerical aperture
NO	Normal order
OPG	Optical parametric generator

PB	Pegg–Barnett
PC	Photon counting
PDC	Parametric down conversion
PJ	Photonic jet
PM	Photomultipliers
POVM	Positive-operator valued measure
PT	Partial transpose
PVM	Projection valued measure
QM	Quantum mechanical
QPD	Quasi probability density
SERS	Surface-enhanced Raman scattering
SG	Susskind and Glogower
SNOM	Scanning near-field optical microscopy
SQL	Standard quantum limit
SS	Squeezed states
SV	Squeezed vacuum
SVA	Slowly varying approximation
WN	White noise

PART I

FUNDAMENTALS OF QUANTUM OPTICS

CHAPTER 1

FUNDAMENTAL PROPERTIES OF COHERENT AND SQUEEZED ELECTROMAGNETIC (EM) STATES

The present chapter reviews the fundamental properties of coherent and squeezed electromagnetic (EM) states, focusing on their quantum characteristics. The creation and annihilation operators are central to the quantization of EM field operators (Glauber 1963a; Glauber 1963b; Loudon 1983; Scully and Zubairy 1997; Walls and Milburn 2008) in relation to the decrease of the quantum noise of squeezed states (SS) relative to that of coherent states (Teich and Saleh 1989). The emphasis in the analysis is placed on SS, which have been obtained and discussed in many experiments (see, e.g., Heidmann et al. 1987; Reynaud et al. 1987; Shelby et al. 1986; Slusher et al. 1986; Wu et al. 1986). An extensive treatment of this topic, supported by numerous calculations, is provided by Loudon and Knight (1987), including the references therein.

Considering a cavity with a volume V, cross-section A and length L along the z-direction. From the boundary conditions, the allowed wave vectors are discrete: $k_m = \pi m / L$. A single cavity mode of the EM field behaves as a simple harmonic oscillator of unit mass. The electric field operator $\hat{E}(z,t)$ is given by:

$$\begin{aligned}\hat{E}(z,t) &= E^{(+)}(z,t) + E^{(-)}(z,t)\\ &= i\sum_m \left(\frac{\hbar\omega_m}{2\varepsilon_0 V}\right)\{\hat{a}(k_m.t)\sin(k_m z) - \hat{a}^\dagger(k_m.t)\sin(k_m z)\},\end{aligned} \tag{1.1}$$

where $\hat{a}$ and $\hat{a}^\dagger$ are the annihilation and creation operators, ω_m is the frequency, and

$$E^{(-)}(z,t) = E^{(+)}(z,t)^{\dagger} . \tag{1.2}$$

The coherent state (CS) $|\alpha\rangle$ can be defined as:

$$|\alpha\rangle = exp\left(-|\alpha|^2/2\right)\sum_{n=0}^{\infty}\frac{\alpha^n}{(n!)^{1/2}}|n\rangle , \tag{1.3}$$

where the number state $|n\rangle$ is given as:

$$|n\rangle = \frac{(\hat{a}^{\dagger})^n}{(n!)^{1/?}}|0\rangle , \tag{1.4}$$

and $|0\rangle$ is the vacuum state. For a CS, the photon number follows a Poisson distribution, where the variance is given by:

$$\left\langle(\nabla n)^2\right\rangle = \langle\alpha|\hat{n}^2|\alpha\rangle - \langle\alpha|\hat{n}|\alpha\rangle^2 = \langle\alpha|\hat{n}|\alpha\rangle , \tag{1.5}$$

where $\hat{n} = \hat{a}^{\dagger}\hat{a}$ is the number operator. We define the quadrature operators (Walls 1983).

$$\begin{aligned}\hat{X} &= \frac{1}{2}(\hat{a} + \hat{a}^{\dagger}) = (\omega/2\hbar)^{1/2}\,\hat{q}\\ \hat{Y} &= \frac{1}{2}(\hat{a} - \hat{a}^{\dagger}) = (2\hbar\omega)^{1/2}\,\hat{p},\end{aligned} \tag{1.6}$$

where $\hat{q}$ and $\hat{p}$ are the space and momentum operators, respectively. For CS, we obtain the following relations:

$$\left\langle(\Delta X)^2\right\rangle = \left\langle(\Delta Y)^2\right\rangle = 1/4 . \tag{1.7}$$

The CS can be generated from the vacuum states by the action of the Glauber unitary displacement operator $\hat{D}(\alpha)$ as:

$$\begin{aligned}&\hat{D}(\alpha) = \exp(\alpha\hat{a}^{\dagger} - \alpha^{*}\hat{a});\ |\alpha\rangle = \hat{D}(\alpha)|0\rangle\\ &\hat{D}^{-1}(\alpha)\hat{a}\hat{D}(\alpha) = \hat{a} + \alpha;\ \hat{D}^{-1}(\alpha)\hat{a}^{\dagger}\hat{D}(\alpha) = \hat{a}^{\dagger} + \alpha^{*}\end{aligned} \tag{1.8}$$

The *two-mode squeezed vacuum* (*SV*) *state* is given by:

$$|r,\theta\rangle = \hat{S}(r,\theta)|0\rangle_1|0\rangle_2$$
$$\hat{S}(r,\theta) = \exp\left(\frac{1}{2}\varsigma^*\hat{a}_1\hat{a}_2 - \frac{1}{2}\varsigma\hat{a}_1{}^\dagger\hat{a}_2{}^\dagger\right);\ \varsigma = re^{i\theta}, \tag{1.9}$$

where $\varsigma = re^{i\theta}$, $\hat{a}_1$ and $\hat{a}_2$ are the annihilation operators for the two modes with frequencies $\Omega_1 = \Omega + \delta$ and $\Omega_2 = \Omega - \delta$. One method for producing this state is by using a parametric amplifier (which will be analyzed in the next chapter), where 2Ω and ϕ represent the frequency and phase of the pump field, respectively. As δ approaches zero, the two-mode SV state can be reduced to a *degenerate one-mode SV state*. The operator $\hat{S}(r,\theta)$ is a unitary operator that transforms the creation and annihilation operators according to the following relations:

$$S^{-1}(\varsigma)\hat{a}_1S(\varsigma) = \hat{a}_1\cosh(r) - a_2{}^\dagger\sinh(r)\exp(i\phi) = \hat{a}_{1,s}$$
$$S^{-1}(\varsigma)\hat{a}_1{}^\dagger S(\varsigma) = \hat{a}_1{}^\dagger\cosh(r) - \hat{a}_2\sinh(r)\exp(-i\phi) = \hat{a}_{1,s}{}^\dagger. \tag{1.10}$$

Similarly, relations for $\hat{a}_2$ and $\hat{a}_2{}^\dagger$, where the new operators are denoted as $\hat{a}_{1,s}$ and $\hat{a}_{1,s}{}^\dagger$, respectively, obey the same *CR*. The *degenerate one-mode SV state* is given by (Walls 1983):

$$\hat{S}(\varsigma)|0\rangle = \exp\left(\frac{1}{2}\varsigma^*\hat{a}^2 - \frac{1}{2}\varsigma\hat{a}^{\dagger 2}\right)|0\rangle$$
$$S^{-1}(\varsigma)\hat{a}S(\varsigma) = \hat{a}\cosh(r) - \hat{a}^\dagger\sinh(r)\exp(i\theta) = \hat{a}_s$$
$$S^{-1}(\varsigma)\hat{a}^\dagger S(\varsigma) = \hat{a}^\dagger\cosh(r) - \hat{a}\ \sinh(r)\exp(-i\theta) = \hat{a}_s{}^\dagger. \tag{1.11}$$

The expectation values for the *quadrature over the degenerate SV state* vanish:

$$\langle 0|\hat{S}^{-1}(r,\theta)\hat{X}\hat{S}(r,\theta)|0\rangle = 0\ ;\ \hat{X} = \left(\frac{\hat{a}+\hat{a}^\dagger}{2}\right)$$
$$\langle 0|\hat{S}^{-1}(r,\theta)\hat{Y}\hat{S}(r,\theta)|0\rangle = 0\ ;\ \hat{Y} = \left(\frac{\hat{a}-\hat{a}^\dagger}{2i}\right), \tag{1.12}$$

as can be verified using Eq. (1.11). The expectation value of the *squared quadrature operator* $\hat{X}^2$ over the degenerate SV state is given by using Eq. (1.11):

$$\begin{aligned}&\langle 0|\hat{S}^{-1}(r,\theta)\hat{X}^2\hat{S}(r,\theta)|0\rangle\\&=\frac{1}{4}\langle 0|\{(\hat{a}+\hat{a}^{\dagger})\cosh r-(e^{i\theta}a+e^{-i\theta}a^{\dagger})\sinh r\}^2|0\rangle\\&=\frac{1}{4}(\cosh^2 r+\sinh^2 r-2\sinh r\cosh r\cos\theta).\end{aligned} \tag{1.13}$$

Assuming that at $t=0$, we have $\theta=0$, obtaining:

$$\langle 0|\hat{S}^{-1}(r,\phi)\hat{X}^2\hat{S}(r,\phi=0)|0\rangle=\frac{1}{4}e^{-2r}. \tag{1.14}$$

We develop, in a similar way, the equation for the *squared quadrature operator* $\hat{Y}^2$ over the degenerate SV state, obtaining:

$$\begin{aligned}&\langle 0|\hat{S}^{-1}(r,\phi)\hat{Y}^2\hat{S}(r,\phi=0)|0\rangle=\frac{1}{4}e^{2r}\\&\left\langle(\Delta X)^2(\Delta Y)^2\right\rangle=\frac{1}{16}.\end{aligned} \tag{1.15}$$

The above calculations reveal that for $\theta=0$ the measurement of quadrature $\hat{X}$ is more accurate at the expense of increasing the noise in the other quadrature $\hat{Y}$. By returning to Eq. (1.13), we find that the quadrature operators in the $X-Y$ plane depends on the angle θ (Fisher et al. 1984). Without the approximation $\theta=0$ more general calculations lead to the equations:

$$\begin{aligned}&\left\langle(\Delta X)^2\right\rangle=\frac{1}{4}\left\{\exp(-2r)\cos^2\frac{1}{2}\theta+\exp(2r)\sin^2\frac{1}{2}\theta\right\};\\&\left\langle(\Delta Y)^2\right\rangle=\frac{1}{4}\left\{\exp(-2r)\sin^2\frac{1}{2}\theta+\exp(2r)\cos^2\frac{1}{2}\theta\right\}.\end{aligned} \tag{1.16}$$

By considering the zeroth-order time dependence $a\to ae^{-i\omega t}$, $a^{\dagger}\to a^{\dagger}e^{i\omega t}$, the CS rotates in the $X-Y$ plane with frequency ω. As a result, it becomes a minimum-uncertainty state only at times $nT/2$, where n is an integer, and *T* is the time period of the EM field. At all other times $\Delta X \Delta Y > 1/4$. By performing interference between the SV state and the two-mode CS, the result can be described as a minimum uncertainty state in a rotating reference frame, where the frame rotates at the angular frequency of the coherent state.

The excited (or displaced) SV state is given by:

$$|\alpha,\varsigma\rangle = \hat{D}(\alpha)\hat{S}(\varsigma)|0\rangle \quad (1.17)$$

The application of Eq. (1.17) is usually based on the conditions that the uncertainties are obtained by the SV state, i.e., by Eq. (1.16), while the light intensity is given mainly by $|\alpha^2|$, resulting in:

$$\langle\alpha,\varsigma|\hat{a}^\dagger\hat{a}|\alpha,\varsigma\rangle = |\alpha|^2 + \sinh^2 r$$
$$\sinh^2 r \,\square\, |\alpha|^2 . \quad (1.18)$$

In this framework, the distribution of the SV state in the complex α plane is represented as an ellipse centered at the origin. The major and minor axes, denoted by ΔX and ΔY are given by Eq. (1.16). The state described in Eq. (1.17) is obtained by displacing the ellipse center by the vector α with the same distribution of uncertainties. Generalization of Eqs. (1.17) and (1.18) to two modes with detailed calculations are given, for example, in (Loudon and Knight 1987). Additional material on squeezed states can be found in other reviews, e.g., Dodonov (2002).

References

Dodonov, V. V. 2002. “Nonclassical 'States in Quantum Optics: A Squeezed Review of the First 75 Years.” Journal of Optics B: Quantum and Semiclassical Optics 4 (1): R1-R33. https://doi.org/10.1088/1464-4266/4/1/201.

Fisher, R. A., M. M. Nieto, and V. D. Sandberg. 1984. “Impossibility of Naively Generalized Squeezed Coherent States.” Physical Review D 29: 1107-1110. https://doi.org/10.1103/PhysRevD.29.1107.

Glauber, R. J. 1963a. “The Quantum Theory of Optical Coherence.” Physical Review 130: 2529–2539. https://doi.org/10.1103/PhysRev.130.2529.

Glauber, R. J. 1963b. "Coherent and Coherent States of the Radiation Field." Physical Review 131: 2766–2788. https://doi.org/10.1103/PhysRev.131.2766.

Heidmann, A., R. J. Horowicz, S. Reynaud, E. Giacobino, E. C. Fabre, and G. Camy. 1987. "Observation of Quantum Noise Reduction on Twin Laser Beams." Physical Review Letters 59: 2555–2558. https://doi.org/10.1103/PhysRevLett.59.2555.

Loudon, R. 1983. The Quantum Theory of Light. Oxford: Clarendon Press

Loudon, R., and P. L. Knight. 1987. "Squeezed Light." Journal of Modern Optics 34: 709–759. https://doi.org/10.1080/09500348714550721.

Reynaud, S., C. Fabre, and E. Giacobino. 1987. "Quantum Fluctuations in a Two-Mode Parametric Oscillator." Journal of the Optical Society of America B 4: 1520–1525. https://doi.org/10.1364/JOSAB.4.001520.

Scully, M. O., and M. S. Zubairy. 1997. Quantum Optics. Cambridge University Press. https://doi.org/10.1017/CBO9780511813993.

Shelby, R. M., M. D. Levenson, S. H. Perlmutter, R. G. DeVoe, and D. F. Walls. 1986. "Broad-Band Parametric De-Amplification of Quantum Noise in an Optical Fiber." Physical Review Letters 57: 691–694. https://doi.org/10.1103/PhysRevLett.57.691.

Slusher, R. E., L. W. Hollberg, B. Yurke, J. C. Mertz, and J. F. Valley. 1986. "Observation of Squeezed States Generated by Four-Wave Mixing in an Optical Cavity." Physical Review Letters 55: 2409–2412. https://doi.org/10.1103/PhysRevLett.55.2409.

Teich, M. C., and B. E. A. Saleh. 1989. "Tutorial: Squeezed States of Light." Quantum Optics 1: 153–191. https://doi.org/10.1088/0954-8998/1/2/006.

Walls, D. F. 1983. "Squeezed States of Light." Nature 306: 141–146. https://doi.org/10.1038/306141a0.

Walls, D. F., and G. J. Milburn. 2008. Quantum Information. In *Quantum Optics,* edited by D. F. Walls and G. J. Milburn. Springer, Berlin, Heidelberg. https://doi.org/10.1007/978-3-540-28574-8_16.

Wu, L. A., H. J. Kimble, J. L. Hall, and H. Wu. 1986. "Generation of Squeezed State by Parametric Down Conversion." Physical Review Letters 57: 2520–2523. https://doi.org/10.1103/PhysRevLett.57.2520.

CHAPTER 2

PHOTON STATISTICS, NONCLASSICAL STATES, AND HOMODYNE DETECTION

Let us consider a photoelectric detector illuminated by a beam of light. The photodetector is based on the photoionization process: an incident photon ionizes an atom of the detector, and the emitted electron is accelerated and amplified to create a pulse. A shutter in front of the phototube controls the length of time for which light falls on the detector. The number of counted photons is recorded, and the experiment is repeated many times for the same opening *T*. The measured statistical distribution $P_m(t)$ contains information about the statistical properties of light.

The standard theory of photodetection is based on the multi-coincidence rate (Kelly and Kleiner 1964):

$$W^{(n)}(t_1 \cdots t_n) = \eta^n Tr\left[\rho \hat{E}^{(-)}(t_1)\cdots \hat{E}^{(-)}(t_n)\hat{E}^{(+)}(t_n)\cdots \hat{E}^{(+)}(t_1)\right], \tag{2.1}$$

where ρ is the density matrix of the field, and η is the quantum efficiency of the detector. The $\hat{E}$ operators are defined, for example, in Eq. (1.1). However, in this case, we assume propagating light rather than light within a cavity.

Instead of expressing the field in terms of spatial modes (which involves performing a Fourier analysis of the spatial variable into wave vectors k_m) we can write it in terms of temporal modes, where we perform a Fourier analysis of the time variable t into discrete frequencies ω_m, where $\omega_m = 2m\pi / T$. The electric field for a plane wave can then be written as (Collett, Loudon, and Gardiner 1987; Huttner and Ben-Aryeh 1990; Collett and Gardiner 1984):

$$\hat{E}(t)=\hat{E}^{(+)}(t)+\hat{E}^{(-)}(t)=\sum_m\left[\frac{\hbar\omega_m}{2\varepsilon_0 cT}\right]^{1/2}[\hat{a}(\omega_m)e^{-i\omega_m t}+H.C.], \tag{2.2}$$

where $\hat{a}(\omega_m)\rightarrow\hat{a}(z,\omega_m)$ and their conjugates form a set of localized creation and annihilation operators (instead of the usual $\hat{a}(k_m,t)$). The cross-section of the beam should be included in the denominator of Eq. (2.2); however, we set it to unity. Following this description, the operator

$$\hat{n}_T=\hat{a}^{\dagger}(\omega_m)\hat{a}(\omega_m), \tag{2.3}$$

is the number operator of the photons of frequency ω_m passing through a certain plane z during a time period T and with a unit cross-sectional area. Here, we focus on the quantization of temporal modes (however, a more general treatment of temporal modes, including the effects of dispersion, is treated in Chapter 3).

The first-order coincidence rate is related to the average photon flux and the quantum efficiency as:

$$W^{(1)}(t)=\left\langle\hat{E}^{(-)}(t)\hat{E}^{(+)}(t)\right\rangle\eta=I(t)\eta. \tag{2.4}$$

where *I(t)* represents the photon flux per unit time and unit area. This quantity is related to the average number of counts during a time interval $[0,T]$ by:

$$\langle\hat{n}\rangle_T=\int_0^T W^{(1)}(t)dt \tag{2.5}$$

The second-order coincidence rate is related to the second factorial moment (Dagenais and Mandel 1978):

$$\langle n^{(2)}\rangle_T=\langle n(n-1)\rangle_T=\int_0^T\int_0^T dt_1 dt_2 W^{(2)}(t_1,t_2)$$
$$W^{(2)}=\eta^2\left\langle\hat{E}^{(-)}(t_1)\hat{E}^{(-)}(t_2)\hat{E}^{(+)}(t_2)\hat{E}^{(+)}(t_1)\right\rangle. \tag{2.6}$$

For stationary light, we apply the following approximations:

$$W^{(2)} = \eta^2 \left\langle \hat{E}^{(-)}(t)\hat{E}^{(-)}(t+\tau)\hat{E}^{(+)}(t+\tau)\hat{E}^{(+)}(t) \right\rangle$$
$$\langle W^{(1)} \rangle = \frac{\langle n \rangle_T}{T}, \tag{2.7}$$

which leads to the second-order normalization condition:

$$g^{(2)}(\tau) = \frac{W^{(2)}(t, t+\tau)}{\langle W^{(1)}(t)\rangle \langle W^{(1)}(t)\rangle}$$
$$= \frac{\eta^2 \left\langle \hat{E}^{(-)}(t)\hat{E}^{(-)}(t+\tau)\hat{E}^{(+)}(t+\tau)\hat{E}^{(+)}(t) \right\rangle}{\langle n \rangle_T^{\ 2} / T^2}. \tag{2.8}$$

Substituting Eq. (2.8) into Eq. (2.6) and changing the variables t_1 and t_2 to t_1 and $\tau = t_2 - t_1$ results in:

$$\langle n(n-1) \rangle = \langle n \rangle_T^{\ 2} / T^2 \int_0^{T-\tau} dt_1 \int_0^T g^{(2)}(\tau) d\tau \quad . \tag{2.9}$$

The upper limit in the integration over t_1 is given by $t_1 = T - \tau$. By rearranging terms, Eq. (2.9) can be rewritten into the following very useful form (Gardiner and Collett 1985; Gardiner and Zoller 2004):

$$\langle n^2 \rangle - \langle n \rangle^2 = \langle n \rangle + \langle n \rangle^2 \left[\frac{2}{T} \int_0^T \left(1 - \frac{\tau}{T} \right) (g^{(2)}(\tau) - 1) d\tau \right]. \tag{2.10}$$

We note that:

$$\langle n \rangle^2 \left[\frac{2}{T} \int_0^T \left(1 - \frac{\tau}{T} \right) (-1) d\tau \right] = -\langle n \rangle^2 . \tag{2.11}$$

This term is inserted on both sides of Eq. (2.10), leading to the following important result: $\langle n^2 \rangle - \langle n \rangle^2 = \langle n \rangle$ that represents the variance for a CS (see Eq. (1.5)). The last term on the right-hand side of Eq. (2.10) shows the deviation of the occupation number distribution from Poisson statistics. This deviation is measured by the Mandel *Q*-factor (Mandel 1986; Mandel 1979):

$$Q = \frac{\langle n^2 \rangle - \langle n \rangle^2}{\langle n \rangle} - 1 \quad . \tag{2.12}$$

For short-times, we apply the approximation $g^{(2)}(\tau) \to g^{(2)}(0)$ in Eq. (2.10), and after performing an integration, we get the following:

$$Q = \langle n \rangle (g^{(2)}(0) - 1) \, , \tag{2.13}$$

where n is the photon number operator, and $g^{(2)}(0)$ is the normalized second-order correlation function at short time intervals. Negative values of Q corresponds to a state whose variance of photon number is less than the mean (i.e., sub-Poisson statistics). Here $-1 \le Q < 0$ presents nonclassical light. The minimum value, $Q = -1$ is obtained for a number state.

In classical light sources, such as thermal light, photons tend to arrive in bunches. However, the phenomenon of antibunching, where photoelectric counts exhibit anti-correlations, was first observed in resonance fluorescence experiments (Kimble, Dagenais, and Mandel 1977; Mandel and Wolf 1995). The number of recorded pulsed pairs as a function of the time delay τ between them showed growth from $\tau = 0$ demonstrating antibunching. While photon bunching has semiclassical interpretations, antibunching can only be understood in terms of a quantized EM field. The antibunching of photons can also be related to the property of the second-order correlation function with $\hat{E}$ operators as a function of τ in Eq. (2.8).

2.1. Photocurrent spectrum

We follow the approach given by Kelly and Kleiner (1964) and define the random function $O(t)$ representing the output of the photomultipliers (PM):

$$O(t) = \sum_{i=0}^{N} eG\delta(t - t_i)Y_i \tag{2.14}$$

where the measurement time T is divided into N intervals: $[t_0,t_1];[t_1,t_2];\cdots[t_{N-1},t_N]$, and N is large enough to ensure that two photoelectrons are not emitted simultaneously. The variable Y_i is a random variable with values of $[t_i,t_{i+1}]:Y_i=1$ if detection occurs and $Y_i=0$ if no detection occurs. Here, $eG\delta(t-t_i)$ represents the intensity of the sharp pulse occurring when a detection event takes place, and G is the gain of the PM.

To obtain the spectrum, we introduce the random variable correlation function for the operator $\hat{O}$ in Eq. (2.14):

$$C(\tau)=\frac{1}{T}\int_0^T O(t)O(t+\tau)d\tau, \tag{2.15}$$

using the relation $\delta(t-t_i)\delta(t-t_j+\tau)\rightarrow\delta(t_i+\tau-t_j)$ we obtain:

$$C(\tau)=\frac{e^2G^2}{T}\sum_{i=0}^{N}\sum_{j=0}^{N}\delta(t_i+\tau-t_j)Y_iY_j \tag{2.16}$$

This leads to two types of terms: for $Y_iY_j=Y_i^2\,(Y_i=Y_j)$ we obtain the average term for $C_1(\tau)$:

$$\langle C_1(\tau)\rangle=\frac{e^2G^2}{T}\left[\int_0^T W^{(1)}(t)\,dt\right]\delta(\tau) \,. \tag{2.17}$$

Applying Eqs. (2.4-2.5), we get:

$$\int_0^T W^{(1)}dt=\langle n\rangle_T=\langle I\rangle T\eta, \tag{2.18}$$

where $\langle I\rangle\eta$ represents the average photon flux. For stationary light, we have:

$$\langle W^{(1)}\rangle=\frac{1}{T}\int_0^T W^{(1)}dt=\langle I\rangle\eta \;;\; \frac{\langle n\rangle_T}{T}=\langle I\rangle\eta \,. \tag{2.19}$$

The average current $\langle i \rangle$ is given by:

$$\langle i \rangle = \eta e G \langle I \rangle \tag{2.20}$$

Thus, Eq. (2.17) can be transformed by substituting Eqs. (2.19-2.20) as:

$$\langle C_1(\tau) \rangle = eG \langle i \rangle \delta(\tau) ; \langle i \rangle = \eta e G \langle I \rangle . \tag{2.21}$$

For $Y_i Y_j \ (i \neq j)$ we evaluate the average of the term $C_2(\tau)$:

$$C_2(\tau) = \frac{e^2 G^2}{T} \int_0^T W^{(2)}(t, t+\tau) dt = e^2 G^2 W^{(2)}(\tau), \tag{2.22}$$

where for stationary light, using Eqs. (2.8) and (2.19-2.20), we obtain:

$$C_2(\tau) = \langle i \rangle^2 g^{(2)}(\tau) . \tag{2.23}$$

By subtracting $\langle i \rangle^2$ from $\langle C(\tau) \rangle = \langle C_1(\tau) \rangle + \langle C_2(\tau) \rangle$ and applying the Fourier transform on $\langle C(\tau) \rangle$ the intensity fluctuation spectrum $N(\omega)$ is obtained as:

$$N^2(\omega) = \frac{eG}{2\pi} \langle i \rangle + \frac{1}{2\pi} \langle i \rangle^2 \int_{-\infty}^{\infty} [g^{(2)}(\tau) - 1] e^{-i\omega\tau} d\tau . \tag{2.24}$$

For small time intervals, we can exchange $[g^{(2)}(\tau) - 1] \rightarrow [g^{(0)}(0) - 1]$. For coherent light, we obtain the standard limit $N^2(\omega) = (eG/2\pi) \langle i \rangle$. Nonclassical light is obtained under the condition $[g^{(0)}(0) - 1] < 1$, where maximal noise reduction occurs in a number state, which satisfies $[g^{(0)}(0) - 1] = -1/n$.

2.2. Homodyne detection of nonclassical light

The general theory of homodyne and heterodyne detection was developed by Huttner and Ben-Aryeh (1989). In the following discussion, we limit our focus to applying this theory to the photocurrent spectrum for analyzing the detection of nonclassical light. In a typical experimental scheme, a coherent local oscillator (*LO*) field and a SS are mixed at a beam splitter (*BS*). The resulting field generates a current $\langle i_1 \rangle$ and $\langle i_2 \rangle$ at the two photomultipliers (*PM1* and *PM2*). These currents are then sent to a frequency analyzer to obtain the noise spectrum. We assume a simple transformation of the fields after splitting at the BS (Huttner and Ben-Aryeh 1988):

$$E_1^{(+)}(t) = S\varepsilon^{(+)}(t) + RE^{(+)}(t)$$
$$E_2^{(+)}(t) = -R\varepsilon^{(+)}(t) + SE^{(+)}(t) \tag{2.25}$$

where $S\,(R)$ is the real transmission (reflection) coefficient, $\varepsilon(t)$ is the SS field before splitting, and $E(t)$ is the coherent *LO* field. We assume steady-state conditions for the fields, such that time dependence enters only via the time differences in the various correlation functions.

To determine the noise in the current, we consider three correlation functions: $g_1^{(2)}(\tau), g_2^{(2)}(\tau), g_{12}(\tau)$ for *PM1* and *PM2*, and the correlation between the two photomultipliers, respectively:

$$g_1^{(2)}(\tau) = \frac{\langle \hat{E}_1^{(-)}(t)\hat{E}_1^{(-)}(t+\tau)\hat{E}_1^{(+)}(t+\tau)\hat{E}_1^{(+)}(t)\rangle}{\langle \hat{E}_1^{(-)}(t)\hat{E}_1^{(+)}(t)\rangle\langle \hat{E}_1^{(-)}(t+\tau)\hat{E}_1^{(+)}(t+\tau)\rangle}, \tag{2.26}$$

and similarly, for $g_2^{(2)}(\tau)$, we have:

$$g_{1,2}^{(2)}(\tau) = \frac{\langle \hat{E}_1^{(-)}(t)\hat{E}_2^{(-)}(t+\tau)\hat{E}_2^{(+)}(t+\tau)\hat{E}_1^{(+)}(t)\rangle}{\langle \hat{E}_1^{(-)}(t)\hat{E}_1^{(+)}(t)\rangle\langle \hat{E}_2^{(-)}(t+\tau)\hat{E}_2^{(+)}(t+\tau)\rangle}. \tag{2.27}$$

Then, by sending $i_1 - i_2$ to a spectrum analyzer, we obtain the noise spectrum:

$$N^2(\omega) = \frac{eG}{2\pi}(\langle i_1 \rangle + \langle i_2 \rangle) + \frac{\langle i_1 \rangle^2}{2\pi}\int_{-\infty}^{\infty}[g_1^{(2)}(\tau) - 1]e^{-i\omega\tau}d\tau$$
$$+\frac{\langle i_2 \rangle^2}{2\pi}\int_{-\infty}^{\infty}[g_2^{(2)}(\tau) - 1]e^{-i\omega\tau}d\tau - 2\frac{\langle i_1 \rangle\langle i_2 \rangle}{2\pi}\int_{-\infty}^{\infty}[g_{1,2}^{(2)}(\tau) - 1]e^{-i\omega\tau}d\tau, \tag{2.28}$$

where $\langle i_1 \rangle$ and $\langle i_2 \rangle$ are the currents at detectors 1 and 2, and G is the gain of both detectors.

To calculate the various correlation functions, we consider the case of a 50/50 $BS\left(R = S = 1/\sqrt{2}\right)$, such that:

$$\langle i_1 \rangle = \langle i_2 \rangle = \frac{\langle i \rangle}{2}. \tag{2.29}$$

Additionally, we specialize in an SV state such that all the odd powers of the SV operators have zero expectation values. We also assume that the light intensity of the coherent state is much larger than the SV state, but the SV state has a large amount of quantum fluctuations. The correlations of Eqs. (2.26) and (2.27) are composed of two distinct types of terms. The first consists of products of four strong fields E operators and the second type includes products of strong fields E operators along with two weakly squeezed fields ε operators. Following the explicit calculations made in (Huttner and Ben-Aryeh 1989), we find that all products of four strong field operators vanish in Eq. (2.28). After carrying out the detailed calculations of Eq. (2.28), we remain with the expression:

$$N^2(\omega) = \frac{eG}{2\pi}\langle i \rangle + \frac{1}{2\pi}\langle i \rangle^2 \int_{-\infty}^{\infty} [\langle g_S{}^{(2)}(\tau)\rangle] e^{-i\omega\tau} d\tau, \tag{2.30}$$

where:

$$\langle g_S{}^{(2)}(\tau)\rangle = \frac{\langle E^{(-)}(t)E^{(-)}(t+\tau)\rangle\langle \varepsilon^{(+)}(t+\tau)\varepsilon^{(+)}(t)\rangle + \langle E^{(-)}(t)E^{(+)}(t+\tau)\rangle\langle \varepsilon^{(-)}(t+\tau)\varepsilon^{(+)}(t)\rangle}{\langle E^{(-)}(t)E^{(+)}(t)\rangle} + C.C., \tag{2.31}$$

and the fields E and ε are of the coherent and squeezed states, respectively, and all the fields are taken in free space before the *BS*. We assume that the use of a laser well above the threshold produces a coherent *LO* field so that its intensity fluctuations can be neglected. Additionally, all our calculations are performed in the rotating frame of the *LO* so that all dependence on the frequency of the *LO* is eliminated. There are various damping mechanisms that will be important in the following calculations.

However, in the present method, we perform the calculations without considering the damping mechanism. Finally, we will discuss how these various damping effects can be incorporated into the final calculations below. Neglecting any damping mechanism, we obtain:

$$\langle E^{(-)}(t)E^{(+)}(t+\tau)\rangle = \langle E^{(-)}(t)E^{(+)}(t)\rangle;$$
$$\langle E^{(-)}(t)E^{(-)}(t)\rangle = e^{2i\phi}\langle E^{(-)}(t)E^{(+)}(t)\rangle, \tag{2.32}$$

where ϕ is the phase of the LO. As we neglect the phase jitter and any damping mechanism, we assume ϕ to be a constant. Under these approximations and for a steady state, substituting Eq. (2.32) in Eq. (2.31) yields:

$$\langle g_S{}^{(2)}(\tau)\rangle = \frac{e^{-2i\theta}\langle \varepsilon^{(+)}(t+\tau)\varepsilon^{(+)}(t)\rangle + \langle \varepsilon^{(-)}(t+\tau)\varepsilon^{(+)}(t)\rangle}{\langle \hat{n}\rangle} \quad +\text{C.C.} \tag{2.33}$$

In the quantization of the fields by Eq. (2.2) (or Eq. (1.1)), certain coefficients appear before the creation and annihilation operators. However, these coefficients cancel out in Eq. (2.33), as they have the same values in the numerator and denominator. Then, by using normalized creation and annihilation operators in Eq. (2.33), we get:

$$\langle g_S{}^{(2)}(\tau)\rangle = \langle 0|\left(\frac{e^{-2i\theta}\hat{a}_S\hat{a}_S}{\langle \hat{n}\rangle} + \frac{\hat{a}_S{}^{\dagger}\hat{a}_S}{\langle \hat{n}\rangle}\right) + H.C.|0\rangle \, . \tag{2.34}$$

As shown in Eq. (2.34), the term involving the products of four strong coherent fields vanishes. One can say that: "*in homodyne detection, classical noise terms are subtracted and we remain with quantum noise*".

We use the relations:

$$S^{-1}(\varsigma)\hat{a}_S S(\varsigma) = \hat{a}\cosh(r) - \hat{a}^{\dagger}\sinh(r)\exp(i\theta)$$
$$S^{-1}(\varsigma)\hat{a}_S^{\dagger} S(\varsigma) = \hat{a}^{\dagger}\cosh(r) - \hat{a}\ \sinh(r)\exp(-i\theta). \tag{2.35}$$

Then, by substituting Eq. (2.35) into Eq. (2.30), we obtain:

$$N^2(\omega) = \frac{eG}{2\pi}\langle i \rangle$$
$$+\frac{1}{2\pi}\langle i \rangle^2 \int_{-\infty}^{\infty} [2\sinh^2 r - \sinh r \cosh r \cos(\theta - 2\phi)] e^{-i\omega\tau} d\tau, \tag{2.36}$$

where $(\theta - 2\phi)$ represents the difference between the SV phase and twice the coherent state phase. Maximal quantum noise reduction can be achieved under the condition $(\theta - 2\phi) = 0$ (which can easily be obtained by adjusting the coherent state phase). Under this condition, we get:

$$2\sinh^2 r - \sinh r \cosh r \cos(\theta - 2\phi) = -1 + e^{-2r} \quad for \quad (2\theta - \phi) = 0, \tag{2.37}$$

which tends to the value of -1 for large r.

It is evident that the damping mechanisms of the laser and SV light will affect the Fourier transform in Eq. (2.36). It is also important to note that the present analysis was conducted for a propagating wave. Usually, it is preferable to produce the *EM* fields in a cavity, as they have well-defined frequencies relative to that of a propagating wave. In this case, one should use "input-output" equations to properly describe the system. Damping mechanisms can also be derived from laser theory (Sargent and Scully 1986; Shapiro, Yuen, and Mata 1979). Rigorous methods describing damping mechanisms were developed by Gardiner and Zoller (2004); however, the use of a phenomenological approach might be more intuitively appealing. The generalization of the Fourier transform in Eq. (2.36), incorporating damping mechanisms, follows the extensive literature published on this subject. In this framework, damping effects can be included to confine noise reduction to a specific frequency band. The range over which this reduction remains effective can also be validated from experimental observations.

It is interesting to note that, in heterodyne detection in a parametric amplifier, the effect of having two frequencies $\Omega \pm \delta$ where 2Ω is the pump frequency, resulting in multiplying $g_{s^{(2)}}(\tau)$ of Eq. (2.30) by the term $\cos(\delta\tau)$, so that for $\tau \to 0$ we return to the present results. Additionally, incorporating this term with the damping mechanisms will influence the Fourier transform in Eq. (2.30).

References

Collett, M. J., and C. W. Gardiner. 1984. "Squeezing of Intracavity and Traveling-Wave Light Fields Produced in Parametric Amplification." *Physical Review A* 30 (3): 1386–1391. https://doi.org/10.1103/PhysRevA.30.1386.

Collett, M. J., R. Loudon, and C. W. Gardiner. 1987. "Quantum Theory of Optical Homodyne and Heterodyne Detection." *Journal of Modern Optics* 34 (6–7): 881–902. https://doi.org/10.1080/09500348714550811.

Dagenais, M., and L. Mandel. 1978. "Investigation of Two-Time Correlations in Photon Emission from a Single Atom." *Physical Review A* 18 (5): 2217–2228. https://doi.org/10.1103/PhysRevA.18.2217.

Gardiner, C. W., and M. J. Collett. 1985. "Input and Output in Damped Quantum Systems: Quantum Stochastic Differential Equations and the Master Equation." *Physical Review A* 31 (6): 3761–3774. https://doi.org/10.1103/PhysRevA.31.3761.

Gardiner, C. W., and P. Zoller. 2004. *Quantum Noise: A Handbook of Markovian and Non-Markovian Stochastic Methods with Applications to Quantum Optics*. Berlin: Springer.

Huttner, B., and Y. Ben-Aryeh. 1988. "Influence of a Beam-Splitter on Photon Statistics." *Physical Review A* 38 (1): 204–211. https://doi.org/10.1103/PhysRevA.38.204.

Huttner, B., and Y. Ben-Aryeh. 1989. "Homodyne and Heterodyne Detection of Squeezed Vacuum States." *Physical Review A* 40 (5): 2479–2484. https://doi.org/10.1103/PhysRevA.40.2479.

Huttner, B., S. Serulnik, and Y. Ben-Aryeh. 1990. "Quantum Analysis of Light Propagation in a Parametric Amplifier." *Physical Review A* 42 (9): 5594–5601. https://doi.org/10.1103/PhysRevA.42.5594.

Kelly, R. L., and W. Kleiner. 1964. "Theory of Electromagnetic Field and Photoelectron Counting." *Physical Review* 136 (4A): A1316–A1326. https://doi.org/10.1103/PhysRev.136.A316.

Kimble, H. J., M. Dagenais, and L. Mandel. 1977. "Photon Antibunching in Resonance Fluorescence." *Physical Review Letters* 39 (11): 691–695. https://doi.org/10.1103/PhysRevLett.39.691.

Mandel, L. 1979. "Sub-Poisson Photon Statistics in Resonance Fluorescence." *Optics Letters* 4 (7): 205–207. https://doi.org/10.1364/OL.4.000205

Mandel, L. 1986. "Non-Classical States of the Electromagnetic Field." *Physica Scripta* T12: 34–42.

https://doi.org/10.1088/0031-8949/1986/T12/005.
Mandel, L., and E. Wolf. 1995. *Optical Coherence and Quantum Optics.* Cambridge: Cambridge University Press.
Sargent, M., M. O. Scully, and W. E. Lamb. 1986. *Laser Physics.* New York: Addison-Wesley.
Shapiro, J. H., H. P. Yuen, and J. A. M. Mata. 1979. "Optical Communications with Two-Photon Coherent States, II: Photoemissive Detection and Structured Receiver Performance." *IEEE Transactions on Information Theory* IT-25 (2): 179–192. https://doi.org/10.1109/TIT.1979.1056033.

CHAPTER 3

SPACE-TIME DESCRIPTION OF ELECTROMAGNETIC WAVE PROPAGATION IN A DISPERSIVE MEDIUM

In quantum field theory, the generator for time evolution is the Hamiltonian $\hat{H}$. For example, the operators $\hat{E}(z,t)$ and $\hat{a}_k$ satisfy the corresponding equations of motion:

$$i\hbar\frac{\partial\hat{E}(z,t)}{\partial t}=\left[\hat{E}(z,t),\hat{H}\right];\quad \frac{d\hat{a}_k(t)}{dt}=\frac{-i}{\hbar}\left[\hat{a}_k(t),\hat{H}\right], \tag{3.1}$$

and the generator for space propagation is the momentum operator $\hat{G}$ (Abram 1987; Abram 1988; Shen 1967):

$$i\hbar\frac{\partial\hat{E}(z,t)}{\partial z}=\left[\hat{E}(z,t),\hat{G}\right];\quad \frac{d\hat{a}_\omega(z)}{dz}=\frac{-i}{\hbar}\left[\hat{a}_\omega(z),\hat{G}\right]. \tag{3.2}$$

These equations hold for any operator and, therefore, provide a description of the time and space evolution of the field.

In the one-dimensional (1D) case of standard field theory, we define

$$\hat{H}=\int_0^L\hat{H}(z)dz \tag{3.3}$$

where $\hat{H}(z)$ is the Hamiltonian density, and L is the quantization length. For transmission into a dielectric, we face a complication: while the frequency remains constant, the energy density of the field changes, such that $\hat{H}$ becomes medium-dependent. To address this issue, Abram (1987; 1988) suggested changing the quantization volume so that $\hat{H}$ remains constant. However, such an approach is not satisfactory when the refractive index is frequency-dependent, as it would require each mode to be quantized with a different volume.

Our approach relies on the fact that, while the energy depends on the medium, the momentum flux is conserved (Shen 1967; Huttner, Serulnik, and Ben-Aryeh 1990). The exact definition of the conserved momentum flux is rooted in the *four-dimensional energy-momentum tensor* $T^{i,k}$ where *i, k* = 0, 1, 2, 3 (Landau and Lifshitz 1985; Jackson 1975; Brevik 1979; Roman 1969; Serulnik and Ben-Aryeh 1991):

$$T^{i,k} = \begin{bmatrix} H & g_x & g_y & g_z \\ S_x & \sigma_{xx} & \sigma_{xy} & \sigma_{xz} \\ S_y & \sigma_{yx} & \sigma_{yy} & \sigma_{yz} \\ S_z & \sigma_{zx} & \sigma_{zy} & \sigma_{zz} \end{bmatrix}. \tag{3.4}$$

The tensor element $T^{00} = H$ represents the energy density, while the vector (S_x, S_y, S_z) is the Poynting vector, representing the *energy flux* in the *x, y,* and *z* directions, respectively. The vector $(g_x.g_y, g_z)$ represents the *density of* the *vectorial momentum* (proportional to $\vec{D}\times\vec{B}$). The $\sigma\, 3\times3$ tensor represents *a flux of momentum* where the columns describe the *x, y, and z* components of the momentum, and the rows refer to *a flux of momentum* in the propagation direction *x, y, and z*, respectively. The use of the vectorial momentum operators S_x, S_y, S_z, i.e., the Poynting vector components are not efficient for wave propagation calculations. In Ben-Aryeh and Serulnik (1991), it has been shown, for example, that the use of $T_{03} = S_z$ does not describe a physically meaningful equation of propagation. The underlying reason is that a proper propagation equation in the *z*-direction requires the use of a 'hyperplane' perpendicular to the *z*-direction, i.e., $dxdycdt$. Therefore, for propagation in the *z*-direction, we should use the *z*-component of the momentum $p^3 = \int T^{33} dxdycdt.$

In Caves and Crouch (1987), it has been suggested to transform the classical Maxwell equations into operator equations and to use commutation relations for the creation and annihilation operators, which are space-dependent. The method was successful in treating problems of propagation, but the physical meaning of these operators remained unclear. This gap was addressed by Huttner, Serulnik, and Ben-Aryeh (1990), who related the use of these operators to the quantization of the flux of photons through a cross-sectional area A over a time interval T (i.e., exchanging the quantization length L with the time T during which the flux of photons is calculated). Following this approach $\hat{a}_\omega^\dagger(z)$ and $\hat{a}_\omega(z)$ serve as the creation and annihilation operators for the flux of photons and $\hat{a}_\omega^\dagger(z)\hat{a}_\omega(z)$ is the number operator for this flux of photons at a distance z. The equations of propagation are given in Eq. (3.2).

There is an important reason for using Eq. (3.2): a monochromatic beam that is transmitted from a free space into a dielectric medium or from one dielectric medium into another preserves the frequency of the beam and not its wave vector. The operators that are used in Eq. (3.1) correspond to EM modes in a cavity with definite $\vec{k}$. In contrast, the operators used in Eq. (3.2) are suited for propagating EM modes with a definite frequency ω. The use of $\hat{a}_k(t)$ or $\hat{a}_\omega(z)$ follows from different forms of second quantization operators for a linear EM harmonic oscillator.

Imoto (Imoto 1989; Imoto, Jeffers, and Loudon 1992) was the first to show that the spatial evolution of the operators can be related to the energy-momentum tensor element T^{33}. He developed the fundamental equations of motion using a Lagrangian formulation in which time and space coordinates are interchanged. In a related study, Ben-Aryeh, Luks, and Peřinová (1992) explored the symmetry between time and space in the classical Poisson brackets, demonstrating the corresponding spatial CR for $\hat{a}(z)$, and $\hat{a}^\dagger(z)$. In a recent article (Horoshko 2022), an extensive analysis of the generator for the spatial evolution of the EM field was described.

In Ben-Aryeh and Serulnik (1991), the momentum operator $\hat{G}(z)$ was calculated by integrating the tensor element T^{33} over $dxdycdt$. The calculation has given the result:

$$\hat{G}(z)=\sum_m \hbar k_m \hat{a}_\omega^\dagger(z)\hat{a}_\omega(z) \quad . \tag{3.5}$$

The general expression for T^{33} as given by (Landau and Lifshitz 1985; Jackson 1975; Brevik 1979; Ben-Aryeh, Luks and Peřinová 1992), is:

$$T^{33} = \frac{1}{2}\left(B_x H_x + B_y H_y - E_z D_z\right) + \frac{1}{2}\left(E_x D_x + E_y D_y - B_z H_z\right). \quad (3.6)$$

For a plane wave propagating in the z-direction (for which $D_z = B_z = 0$) we get that T^{33} is *always positive*. Thus, Eq. (3.5) should be replaced with a more general form, as derived in (Toren and Ben-Aryeh 1994):

$$\hat{G}(z) = \sum_m \hbar \left|k_m\right| \hat{a}_\omega^\dagger(z) \hat{a}_\omega(z). \quad (3.7)$$

Eq. (3.7) is central for treating coupling between EM modes propagating in opposite directions. For example, this equation was used for the quantum mechanical (QM) treatment of four-wave mixing (Peřina and Křepelka 1992). It was also used in (Toren and Ben-Aryeh 1994) for *quantization* of the classical distributed feedback lasers (Yariv 1983; Yariv and Yeh 1984). Additionally, the quantum statistics for lossless linear couplers were analyzed using this approach (Peřinová et al. 1991). Broader fundamental concepts in quantum optics were described and analyzed in the books by Jan Perina (Peřina, Hradil, and Jurčo 1994; Peřina 1991).

We show here the use of the momentum operator formalism in the treatment of both co-directional and contra-directional propagation of EM waves. The classical analysis of co-directional and contra-directional propagation is based on the coupled-wave theory (Yariv 1983; Yariv and Yeh 1984) for EM modes propagating in a dielectric medium with a specified perturbation. The perturbation couples different modes, leading to the transfer of energy and momentum between them. This perturbation is sufficiently small so that one can use it in theoretical analysis by applying the basis of unperturbed modes. Since the perturbation is small, it is possible to neglect the second derivatives of the amplitudes by applying the slowly varying approximation (SVA). Under these conditions, the classical treatment of two coupled modes leads to two coupled equations depending on the *momentum* phase-matching parameter $\Delta\beta$:

$$\Delta\beta = \beta_2 - \beta_1 - m2\pi / \Lambda. \quad (3.8)$$

Here m is an integer, β_1 and β_2 are the components of the wave vectors of the two modes in the direction of propagation *z*. Λ is the 'wavelength' of the spatial periodic change in the refractive index (considered as a perturbation in the dielectric medium).

In the following sections, we develop the *quantization* of the co-directional and contra-directional systems, which give equivalent results to the classical ones (Yariv 1983; Yariv and Yeh 1984), for these linear systems. Our analysis demonstrates that both the Hamiltonian and the momentum operators should be used to fully account for the conservation of energy and momentum, respectively. In the present analysis, the conventional CRs are not changed (Toren and Ben-Aryeh 1994), and Eq. (3.7) is used for both co-directional and contra-directional coupling.

3.1. Co-directional coupling

In the present section, we present a general QM analysis of a co-directional coupled system by using both momentum and Hamiltonian operators. The zero-order momentum and Hamiltonian are given, respectively, by:

$$\hat{G}_0 = \hbar\left[\beta_1 \hat{a}_1^\dagger \hat{a}_1 + \beta_2 \hat{a}_2^\dagger \hat{a}_2\right]; \quad \beta_1, \beta_2 > 0;$$
$$\hat{H}_0 = \hbar\left[\omega_1 \hat{a}_1^\dagger \hat{a}_1 + \omega_2 \hat{a}_2^\dagger \hat{a}_2\right]; \quad \omega_1, \omega_2 > 0. \tag{3.9}$$

The zero-order time dependence obtained from the Hamiltonian is expressed as:

$$\frac{d\hat{a}_1}{dt} = -\frac{i}{\hbar}[\hat{a}_1, H_0] \Rightarrow \hat{a}_1(0)e^{-i\omega t};$$
$$\frac{d\hat{a}_2}{dt} = -\frac{i}{\hbar}[\hat{a}_2, H_0] \Rightarrow \hat{a}_2(0)e^{-i\omega t}. \tag{3.10}$$

Similarly, the zero-order space dependence obtained from the zero-order momentum operator is given by:

$$\frac{d\hat{a}_1}{dz} = \frac{i}{\hbar}\left[\hat{a}_1, \hat{G}_0\right] \Rightarrow \hat{a}_1(0)e^{i\beta_1 t};$$
$$\frac{d\hat{a}_2}{dz} = \frac{i}{\hbar}\left[\hat{a}_2, \hat{G}_0\right] \Rightarrow \hat{a}_2(0)e^{i\beta_2 t}. \tag{3.11}$$

According to Eqs. (3.9-3.11), $\hat{a}_1$ and $\hat{a}_2$ represent (in the zero-order) modes that are propagating in the positive direction of the z-axis. The zero-order Hamiltonian is unaffected by the perturbation, and the energy flux, which is related to S_z in Eq. (3.4) is given, in accordance with Eq. (3.8) by:

$$\left\langle \hat{H}_0 \right\rangle = \hbar\left[\omega_1 \langle \hat{n}_1 \rangle + \omega_2 \langle \hat{n}_2 \rangle\right]. \tag{3.12}$$

This result is in agreement with the classical description (Yariv 1983; Yariv and Yeh 1984).

The space-dependent perturbation enters the analysis through the momentum operator, including interaction terms, of the form $\tilde{\kappa}\hat{a}_1\hat{a}_2^\dagger + H.C.$ Here

$$\tilde{\kappa} = \kappa \exp\left(-\frac{im2\pi}{\Lambda} z\right), \tag{3.13}$$

in agreement with the classical analysis (Yariv 1983; Yariv and Yeh 1984). The total momentum of the system is given by:

$$\hat{G} = \hbar\left[\beta_1 \hat{a}_1^\dagger \hat{a}_1 + \beta_2 \hat{a}_2^\dagger \hat{a}_2 + \tilde{\kappa}\hat{a}_1\hat{a}_2^\dagger + \tilde{\kappa}^{\square}\hat{a}_1^\dagger \hat{a}_2\right]. \tag{3.14}$$

The equations of motion obtained from Eq. (3.14) are given by:

$$\frac{d\hat{a}_1}{dz} = \frac{i}{\hbar}[\hat{a}_1, G] = i\beta_1\hat{a}_1 + i\tilde{\kappa}^{\square}\hat{a}_2;$$
$$\frac{d\hat{a}_2}{dt} = \frac{i}{\hbar}[\hat{a}_2, G] \Rightarrow i\beta_2\hat{a}_2 + i\tilde{\kappa}\hat{a}_1. \tag{3.15}$$

We define:

$$\hat{a}_1 = \hat{A}_1(z)\exp(i\beta_1 z); \quad \hat{a}_2 = \hat{A}_2(z)\exp(i\beta_2 z) \tag{3.16}$$

Then, by substituting Eqs. (3.16) into Eq. (3,15), we get:

$$\frac{d\hat{A}_1(z)}{dz}\exp(i\beta_1 z) = i\tilde{\kappa}\hat{a}\hat{A}_2(z)\exp(i\beta_2 z) \tag{3.17}$$

By substituting Eq. (3.13) into Eq. (3.17) and rearranging terms, we get:

$$\frac{d\hat{A}_1(z)}{dz} = i\kappa\hat{A}_2(z)\exp[i(\beta_2 - \beta_1 - m2\pi/\Lambda)z]$$
$$\equiv i\kappa\hat{A}_2(z)\exp[i\Delta\beta z]; \quad \Delta\beta = \beta_2 - \beta_1 - m2\pi/\Lambda, \tag{3.18}$$

where $\Delta\beta$ is the *momentum* phase-matching parameter defined in Eq. (3.8).

Using a similar derivation for $\frac{d\hat{A}_2(z)}{dz}$, we get:

$$\frac{d\hat{A}_2(z)}{dz} = i\kappa\hat{A}_1(z)\exp(-i\Delta\beta z) \tag{3.19}$$

The classical amplitudes $A_1(z)$ and $A_2(z)$ (Yariv 1983; Yariv and Yeh 1984) obey the same equations as derived here for the QM operators under the correspondence $\hat{A}_1(z)_{QM} \rightarrow A_1(z)_{clas}; \quad \hat{A}_2(z)_{QM} \rightarrow A_2(z)_{clas}$.

3.2. Contra-directional coupling

In the present section, we develop QM equations that relate the output momentum field operators to the input field operators for contra-directional coupling. We focus on the coupling induced by a spatial periodic dependence in the index of refraction of the medium of two EM modes whose energy flux is propagating in *opposite directions*. For this case, the zero-order Hamiltonian is given by:

$$\hat{H}_0 = \hbar\left[\omega_1 \hat{a}_1{}^\dagger \hat{a}_1 - \omega_2 \hat{a}_2{}^\dagger \hat{a}\ \right];$$
$$\left\langle \hat{H}_0 \right\rangle = \hbar\left[\omega_1 \left\langle \hat{n}_1 \right\rangle - \omega_2 \left\langle \hat{n}_2 \right\rangle\right], \tag{3.20}$$

in agreement with the classical description. The perturbation is introduced only through the momentum operators. Using Eq. (3.7), we find that the zero-order momentum in the contra-directional coupling is equal to that of the co-directional coupling. Consequently, the total momentum operator, given in Eq. (3.14), is the same for the two systems. However, the use of the absolute value for the momentum in Eq. (3.7) introduces a problem in the QM classical analog for contra-directional coupling. This issue was solved by using CR $[\hat{a}_2(z), \hat{a}_2{}^\dagger(z)] = -1$ (Ben-Aryeh, Luks, and Peřinová 1992). As shown in Toren and Ben-Aryeh (1994), the conventional CR

$[\hat{a}_2(z), \hat{a}_2{}^\dagger(z)] = 1$ should be applied. Therefore, following the above considerations, the momentum perturbation in contra-directional coupling is equivalent to that of the co-directional coupling, and all the perturbation analyses made in the previous section also remain valid to that of the contra-directional coupling. This procedure yields results that are in full agreement with the classical analysis (Yariv 1983; Yariv and Yeh 1984).

The difference between co-directional coupling and contra-directional coupling follows from the difference between Eq. (3.20) and Eq. (3.9), while the flux of energy obtained by S_z (representing the Poynting vector) has the same direction for the two EM modes in the co-directional coupling, in contra-directional coupling, the energy flux is propagating in opposite directions for the two modes.

3.3. Quantization in a linear dielectric medium

The field operators change according to the refractive index of the medium $n(\omega_m)$ as shown by Abram (1987):

$$\hat{E}(z,t)=\sum_m\left[\frac{\hbar\omega_m}{2\varepsilon_0 cT\eta(\omega_m)}\right]^{1/2}\left[\hat{a}(z,\omega_m)\exp(-i\omega_m t+H.C.)\right];$$

$$\hat{B}(z,t)=\sum_m\left[\frac{\hbar\omega_m\eta(\omega_m)}{2\varepsilon_0 cT}\right]^{1/2}\left[\hat{a}(z,\omega_m)\exp(-i\omega_m t+H.C.)\right];$$

$$\hat{D}(z,t)=\sum_m\left[\frac{\hbar\omega_m\varepsilon_0}{2cT\eta(\omega_m)}\right]^{1/2}\eta(\omega_m)^2\left[\hat{a}(z,\omega_m)\exp(-i\omega_m t+H.C.)\right]. \tag{3.21}$$

The z-dependence of the operator $\hat{a}(z,\omega_i)$ is given by:

$$\hat{a}(z,\omega_i)=\hat{a}(0,\omega_i)\exp(ik_i z) \tag{3.22}$$

where $k_i=\eta(\omega_i)\omega_i/c$ is the wave vector in the medium. The CR at different points is thus given by:

$$\left[\hat{a}(z,\omega_i),\hat{a}^\dagger(z',\omega_j)\right]=\delta_{i,j}\exp\left[ik_i(z-z')\right]. \tag{3.23}$$

Small, non-linear changes in the refractive index can be incorporated using perturbation theory.

References

Abram, I. 1987. “Quantum Theory of Light Propagation: Linear Medium.” *Physical Review A* 35 (10): 4661–4672. https://doi.org/10.1103/PhysRevA.35.4661.

Abram, I. 1988. “Propagation of Nonclassical Light.” In *Photons and Quantum Fluctuations*, edited by E. R. Pike and H. Walther, 1st ed. Bristol: Hilger. https://doi.org/10.1201/9781003069539-9.

Ben-Aryeh, Y., and S. Serulnik. 1991. “The Quantum Treatment of Propagation in Non-Linear Optical Media by the Use of Temporal Modes.” *Physics Letters A* 155 (9): 473–478. https://doi.org/10.1016/0375-9601(91)90650-W.

Ben-Aryeh, Y., A. Luks, and V. Peřinová. 1992. “The Concept of Equal Space Commutators in Quantum Optics.” *Physics Letters A* 165 (1): 19–23. https://doi.org/10.1016/0375-9601(92)91047-U.

Brevik, I. 1979. "Experiments in Phenomenological Electrodynamics and the Electromagnetic Energy-Momentum Tensor." *Physics Reports* 52 (3): 133–201. https://doi.org/10.1016/0370-1573(79)90074-7.

Caves, C. M., and D. D. Crouch. 1987. "Quantum Wideband Traveling-Wave Analysis of a Degenerate Parametric Amplifier." *Journal of the Optical Society of America B* 4 (10): 1535–1545. https://doi.org/10.1364/JOSAB.4.001535.

Horoshko, D. B. 2022. "Generator of the Spatial Evolution of the Electromagnetic Field." *Physical Review A* 105 (1): 013708. https://doi.org/10.1103/PhysRevA.105.013708.

Huttner, B., S. Serulnik, and Y. Ben-Aryeh. 1990. "Quantum Analysis of Light Propagation in a Parametric Amplifier." *Physical Review A* 42 (9): 5594–5601. https://doi.org/10.1103/PhysRevA.42.5594.

Imoto, N. 1989. "Quantum Non-Demolition Measurements." In *Proceedings of the 3rd International Symposium on the Foundations of Quantum Mechanics*, edited by S. Kobayashi, H. Egawa, Y. Muramaya, and S. Nomura, 306. Tokyo: Physical Society of Japan.

Imoto, N., J. R. Jeffers, and R. Loudon. 1992. "Quantum mechanical treatment of a propagating optical beam." In *Quantum Measurements in Optics*, pp. 295-311. Boston, MA: Springer US.

Jackson, J. D. 1975. *Classical Electrodynamics*. 2nd ed. New York: Wiley.

Landau, L. D., and E. M. Lifshitz. 1985. *The Classical Theory of Fields*. 4th ed. Oxford: Pergamon Press.

Peřina, J., and J. Křepelka. 1992. "Forward and Backward Four-Wave Mixing of Non-Classical Light with Pump Depletion." *Journal of Modern Optics* 39 (11): 2405–2419. https://doi.org/10.1080/09500349214552441.

Peřina, J., Z. Hradil, and B. Jurčo. 1994. *Quantum Optics and Fundamentals of Physics*. Springer.

Peřina, J. 1991. *Quantum Statistics of Linear and Non-Linear Optics*. Dordrecht: Springer.

Peřinová, V., A. Lukš, J. Křepelka, C. Sibilia, and M. Bertolotti, 1991. "Quantum Statistics of Light in a Lossless Linear Coupler." *Journal of Modern Optics* 38 (12): 2429–2444. https://doi.org/10.1080/09500349114552581.

Roman, P. 1969. *Introduction to Quantum Field Theory*. New York: Wiley.

Serulnik, S., and Y. Ben-Aryeh. 1991. "Space-Time Description of Propagation in Nonlinear Dielectric Media." *Quantum Optics: Journal of the European Optical Society Part B* 3 (1): 63–71.

https://doi.org/10.1088/0954-8998/3/1/006.
Shen, Y. R. 1967. "Quantum Statistics of Nonlinear Optics." *Physical Review* 155 (3): 921–933. https://doi.org/10.1103/PhysRev.155.921.
Toren, M., and Y. Ben-Aryeh. 1994. "The Problem of Propagation in Quantum Optics, with Applications to Amplification, Coupling of EM Modes and Distributed Feedback Lasers." *Quantum Optics: Journal of the European Optical Society Part B* 6 (5): 425–437. https://doi.org/10.1088/0954-8998/6/5/006.
Yariv, A. 1983. *Quantum Electronics*. 3rd ed. New York: Holt, Rinehart and Winston.
Yariv, A., and P. Yeh. 1984. *Optical Waves in Crystals: Propagation and Control of Laser Radiation*. New York: Wiley.

CHAPTER 4

SQUEEZING AND PHOTON CORRELATIONS IN A PARAMETRIC AMPLIFIER

A resonant parametric amplifier inside a cavity is an effective device for producing squeezed states (SS). In this set-up, a strong, 'undepleted' pump photon at frequency 2Ω is converted, under resonance conditions, into a pair of photons as:

$$2\Omega = \Omega_1^{\ 0} + \Omega_2^{\ 0} , \tag{4.1}$$

where $\Omega_1^{\ 0}$ and $\Omega_2^{\ 0}$ are the cavity frequencies. For a single-mode cavity, we assume $\Omega_1^{\ 0} = \Omega_2^{\ 0}$, and deviations from resonance are considered (Zahler and Ben-Aryeh 1991). A quantum-mechanical description of such a scheme was first developed by Carmichael, Milburn, and Walls (1984) for the degenerate cavity-mode case (where $\Omega_1^{\ 0} = \Omega_2^{\ 0}$). In their analysis, the authors consider the EM pump field at 2Ω as a classical driving field and neglect the damping mechanism. More general treatments that include fluctuation-dissipation mechanisms were developed in other works (Kubo 1966). In the following analysis, we concentrate on the pure detuned vacuum SS and its photon number distribution, which is affected by the detuning parameter. We use the Lie group representations technique, where operators act on the vacuum in normal order (NO) form (Gilmore 1974). Oscillations observed in the photon number distribution of resonantly excited coherent-squeezed states have been interpreted as interference effects, explained through Wigner functions (Schleich, Walls, and Wheeler 1988) and through interference between phase space representations (Kim, de Oliveira, and Knight 1989).

4.1. Photon number distribution in detuned squeezed vacuum state

A simple model for degenerate parametric down conversion (PDC) is described by the Hamiltonian (Mollow and Glauber 1967):

$$\hat{H}(t) = \hbar\Omega_0 \hat{a}^\dagger \hat{a} - i\frac{\hbar\kappa\varepsilon}{2}\left(\hat{a}^2 \exp\left[i(2\Omega t + \phi)\right] - H.c\right);$$
$$|\Delta\Omega| = |\Omega - \Omega_0| \ll \Omega. \tag{4.2}$$

This model involves several classic parameters: ε - the constant amplitude of the pump EM assumed to remain un-depleted, ϕ - the phase difference between the pump field and the degenerate cavity EM mode, and κ - describing an effective macroscopic nonlinear coupling strength. In this framework, 2Ω and Ω_0 are the pump and cavity mode frequencies, respectively. It is important to note that in the present model, damping mechanisms are neglected.

The photons emitted from the cavity have a half-pump frequency Ω, in accordance with energy conservation. Due to momentum conservation (phase-matching conditions), the corresponding wave vectors satisfy the relation:

$$\vec{k}_p = \vec{k}_s + \vec{k}_i, \tag{4.3}$$

where $\vec{k}_p$, $\vec{k}_s$ and $\vec{k}_i$ are the wave vectors of the pump, the "signal", and the "idler", respectively. Such a "light cone" is reduced, in certain crystals, to an almost parallel propagation by a certain adjustment of the nonlinear uniaxial crystal optical axis relative to the pump direction. Under this assumption, in the present degenerate case, only single-frequency down-converted photons are assumed to leave the cavity within a direction parallel to the pump.

Following the ordinary method for nearly resonant interaction, we define a unitary operator:

$$U(t) = \exp\left(-i\Omega\hat{a}^\dagger \hat{a} t\right), \tag{4.4}$$

so that the Hamiltonian in the rotated picture becomes time-independent:

$$\hat{H} = \hat{U}^{\dagger}(t)\hat{H}(t)\hat{U}(t) = (\Omega - \Omega_0)\hat{a}^{\dagger}\hat{a} - i\hbar\frac{\kappa\varepsilon}{2}\left(a^2 e^{i\phi} - H.c\right). \tag{4.5}$$

Here, the detuning frequency and the squeezing frequency are defined, respectively, as:

$$\Delta\Omega = \Omega - \Omega_0; \quad s = \kappa\varepsilon. \tag{4.6}$$

We also define the following dimensionless parameters: the squeezing parameter τ, the detuning parameter δ and the threshold parameter ξ, given as:

$$\tau = st; \quad \delta = \frac{\Delta\Omega}{s} = \frac{\Omega - \Omega_0}{s}; \quad \xi = \left(1 - \delta^2\right)^{1/2}. \tag{4.7}$$

The detuned squeezed vacuum state $|DVSV\rangle$, is then given by the time evolution operator $\hat{S}_d$ as:

$$|DVSV\rangle = \hat{S}_d|0\rangle = \exp\left[i\delta\tau\hat{a}^{\dagger}\hat{a} - i(\tau/2)\left(a^2 e^{i\phi} - a^{\dagger 2}e^{-i\phi}\right)\right]. \tag{4.8}$$

Since we aim to apply this operator to the vacuum state $|0\rangle$, it is worthwhile to rearrange it in an NO form. To do so, we define:

$$\hat{x}_1 = \hat{a}^{\dagger}\hat{a} + \frac{1}{2} \quad ; \quad \hat{x}_2 = \hat{a}^2 \quad ; \quad \hat{x}_3 = \hat{a}^{\dagger 2}. \tag{4.9}$$

These operators are elements of a closed Lie algebra, generating the group SL(2R). They obey the following CR:

$$[\hat{x}_1, \hat{x}_2] = -2\hat{x}_2 \quad ; \quad [\hat{x}_1, \hat{x}_3] = 2\hat{x}_3 \quad ; \quad [\hat{x}_2, \hat{x}_3] = 4\hat{x}_1, \tag{4.10}$$

and their faithful 2×2 matrix representations may be given as:

$$\hat{x}_1 \to X_1 = \begin{bmatrix} 1 & 0 \\ 0 & -1 \end{bmatrix}; \quad \hat{x}_2 \to X_2 = \begin{bmatrix} 0 & 0 \\ -2 & 0 \end{bmatrix}; \quad \hat{x}_3 \to X_3 = \begin{bmatrix} 0 & 2 \\ 0 & 0 \end{bmatrix}. \tag{4.11}$$

Using the definitions from Eq. (4.9), the normal-ordered form of $\hat{S}_d$ is written as:

$$\hat{S}_D = \exp(\eta\hat{x}_3)\exp(\beta\hat{x}_1)\exp(\gamma\hat{x}_2). \tag{4.12}$$

Our purpose now is to find the relations between the NO parameters η, β, and γ, and the physical parameters, i.e., δ and τ. A General Baker-Campbell-Hausdorff (BCH) theorem (Gilmore 1974) enables us to replace the Lie algebra operators with their matrix representations in order to obtain the desired relations. The BCH equation is given as:

$$\begin{aligned}&\exp(\eta X_3)\exp(\beta X_1)\exp(\gamma X_2)\\&=\exp\left[i\delta\tau X_1 - i(\tau/2)\left(X_2 e^{i\phi} - X_3 e^{-i\phi}\right)\right].\end{aligned} \tag{4.13}$$

We find η, β and γ by comparing the calculated matrix elements on both sides of this equation. After straightforward algebra, we get:

$$\begin{aligned}&e^{\beta} = \frac{1}{\cosh(\xi\tau) - i(\delta/\xi)\sinh(\xi\tau)};\\&\eta = \frac{1}{2\xi}e^{-i\phi}\sinh(\xi\tau)e^{\beta}; \quad \gamma = -\eta e^{2it}.\end{aligned} \tag{4.14}$$

Now, since $\hat{a}|0\rangle = 0$, the action of the NO operator $\hat{S}_D$ from Eq. (4.12) on the vacuum state leads to the simple expression:

$$|DVSV\rangle = \hat{S}_d|0\rangle = \exp(\beta/2)\exp\left(\eta a^{\dagger 2}\right)|0\rangle. \tag{4.15}$$

By substituting η and e^{β} from Eq. (4.14) into Eq. (4.15) and carrying out straightforward algebra, we get:

$$\begin{aligned}&|DVSV\rangle =\\&\frac{\left[(2n!)^{1/2}/n!\right](1/2)^n\left[\sinh(\xi\tau)/\xi\right]^n}{\left[\cosh(\xi\tau) - i(\delta/\xi)\sinh(\xi\tau)\right]^{n+1/2}}\exp\left[-in(\phi + 2\Omega t)\right]|2n\rangle.\end{aligned} \tag{4.16}$$

The photon pair probabilities are given by:

$$P_{2n}=(1/2)^{2n}\left[(2n!)/n!^2\right]\frac{\left[\sinh^2(\xi\tau)/\xi^2\right]^n}{\left[\cosh^2(\xi\tau)+(\delta/\xi)^2\sinh^2(\xi\tau)\right]^{n+1/2}}. \tag{4.17}$$

Here, we obtain the effect of deviation from resonance on the photon number distribution. Under the condition of exact resonance, the following relations hold:

$$\delta=\frac{\Delta\Omega}{s}=0 \quad ; \quad \xi=\left(1-\delta^2\right)^{1/2}=1. \tag{4.18}$$

In this case, the probability is given by (Kim, de Oliveira, and Knight 1989):

$$P_{2n}=(1/2)^{2n}\left[(2n!)/n!^2\right]\frac{\tanh^2(\tau)}{\cosh\tau}. \tag{4.19}$$

(Note: quite often, the notation in the literature is different as $\tau\rightarrow r$).

The method described above has also been applied to more complex systems. For example, Zahler and Ben-Aryeh (1991) applied it to the detuned excited squeezed state, and in a subsequent work (Zahler and Ben-Aryeh 1992) to the two-mode vacuum and excited squeezed states. These results are complicated and more general than the common ones, as they consider the effects of deviation from resonance.

4.2. Photon correlations in a parametric amplifier

Important two-photon correlations between signal and idler produced by a parametric amplifier were investigated by Mandel and his colleagues (e.g., Ghosh and Mandel 1987; Friberg, Hong, and Mandel 1985; Hong, Ou, and Mandel 1987; Ou and Mandel 1988). Fourth-order interference between the signal and idler fields was obtained to measure the joint probability ($P_{1,2}(x_1,x_2)\delta x_1\delta x_2$) of detecting the two photons at x_1 and x_2 within δx_1 and δx_2 , respectively. This probability is obtained by the fourth-order correlation function (Ghosh and Mandel 1987):

$$P_{1,2}(x_1,x_2)\delta x_1\delta x_2 = K\left\langle \hat{E}^{(-)}(x_1)\hat{E}^{(-)}(x_2)\hat{E}^{(+)}(x_2)\hat{E}^{(+)}(x_1)\right\rangle, \tag{4.20}$$

where K is a constant that depends on the geometry of the system. The temporal spread between the two photons produced in the process of spontaneous parametric down-conversion was found to be independent of the pump's coherence time and the propagation time through the crystal (Friberg, Hong, and Mandel 1985). Signal and idler with different frequencies, ω_1 and ω_2, were produced, mixed together, and directed to two photodetectors centered at ω_1 and ω_2, respectively (Ou and Mandel 1988). The coincidence measurements exhibited a modulation of the form $\cos(\omega_1 - \omega_2)\tau$, where $c\tau$ is the path difference. The quantum mechanical analysis of this system employing a 50:50 BS led to the result (Ou and Mandel 1988):

$$P_{1,2} = C\exp\left[(\omega_1+\omega_2-\omega_0)/2\sigma^2\right]\left[\left(1-\exp\left(-\sigma^2\delta\tau^2/2\right)\cos(\omega_1-\omega_2)\tau\right)\right]. \tag{4.21}$$

The first exponential on the right side of Eq. (4.25) represents the effect of deviation from resonance, where ω_0, is the pump frequency and σ is the small frequency width of the filters placed before the detectors. C is a constant related to the geometric properties of the system (Ou and Mandel 1988).

Certain properties of two-photon entangled wave packets produced by a parametric amplifier have been described by optical Wigner functions (Dragoman 1997; Ben-Aryeh, Shih, and Rubin 1999). However, a complete description should be related to the QM fourth-order correlation function, as given above. A pair of signal and idler photons, degenerate in frequency and propagation direction, was used to produce an Einstein-Podolsky-Rosen (EPR)-Bohm coincidence detection at spatially separated detectors, where polarization correlations characteristic of the EPR-Bohm type were observed (Shih and Sergienko 1994; Shih and Alley 1988; Rubin et al. 1994). Violation of Bell's inequality was reported (Shih and Alley 1988), along with optical imaging (Pittman et al. 1995). (Note: The topic of Bell inequalities will be treated in Chapter 12).

4.3. The power of an optical parametric generator by discrete mode summation

An analytical model for calculating the output power of an optical parametric generator (OPG) was previously developed (Pittman et al. 1995). The common approach for describing propagation in such systems is to use a Hamiltonian for the volume and relate time and space by the velocity of propagation. Since, in our case, the propagating waves (pump, signal, and idler) do not share the same velocity, this assumption is no longer valid. The two-mode operator for our system is given by (Huttner, Serulnik, and Ben-Aryeh 1990; Toren and Ben-Aryeh 1994; Byer and Tang 1975):

$$G = \hbar g\left(\hat{a}_s^{\dagger} a_i^{\dagger} \exp(-i\Delta kz) + H.c.\right) \tag{4.22}$$

where $\Delta\vec{k} = \vec{k}_p - \vec{k}_s - \vec{k}_i$ is a vector assumed to lie along the propagation direction z. The parameters k_p, k_s, and, k_i represent the pump, signal, and idler, respectively. The parameter $\vec{k}_P = (2\pi / \Lambda)\hat{z}$ is the phase-matched reciprocal vector, where Λ is modulation parameter of the nonlinear coefficient. The gain coefficient and its dependence on system parameters have been studied (Acco et al. 2011; Byer and Tang 1975) to derive the explicit equations of motion for propagation:

$$\begin{aligned} \frac{d\hat{a}_s}{dz} &= ig\hat{a}_i^{\dagger} \exp(-i\Delta kz); \\ \frac{d\hat{a}_i^{\dagger}}{dz} &= -ig\hat{a}_s \exp(i\Delta kz). \end{aligned} \tag{4.23}$$

After solving these equations, calculations were performed (Acco et al. 2011) for $\langle \hat{a}_s^{\dagger}\hat{a}_s \rangle$, from which the photon number for each mode was found:

$$n_s(L) = n_i(L) = \left(\frac{g}{b}\right)^2 \sinh^2(bL). \tag{4.24}$$

These calculations were made at a distance $z = L$ along the crystal, where b is the parametric gain term (that considers the effect of phase mismatch):

$$b = \sqrt{g^2 - \left(\frac{\Delta k}{2}\right)^2}. \tag{4.25}$$

The model predictions are in excellent agreement with the measured signal power of OPGs based on 25 and 35 mm periodically poled $LiNbO_3$ crystal. It was found that summation over discrete modes gives more accurate results than the traditional method of continuous integration over all parameters.

4.4. Phase estimation by photon counting

Photon counting (PC) measurements were analyzed for obtaining a classical phase parameter in a linear Mach-Zehnder interferometer by using phase estimation theory (Ben-Aryeh 2012). A detailed analysis was carried out for four cases: a) Coherent states inserted into the interferometer; b) A Fock number state inserted in one input port and the vacuum into the other input port; c) A coherent state inserted in one input port and squeezed vacuum in the other; and, d) Exchanging the first beam-splitter by a nonlinear-system which inserts a NOON state into the interferometer, followed by PC for parity measurements. The properties of PC for obtaining minimal phase uncertainties were discussed.

References

Acco, S., A. Arie, Y. Ben-Aryeh, M. Katz, and P. Blau. 2011. "Modeling the Power of an Optical Parametric Generator by Discrete Mode Summation." *Applied Physics B* 103 (1): 89–96. https://doi.org/10.1007/s00340-011-4464-9.

Ben-Aryeh, Y. 2012. "Phase Estimation by Photon Counting Measurements in the Output of a Linear Mach–Zehnder Interferometer." *Journal of the Optical Society of America B* 29 (10): 2754–2759. https://doi.org/10.1364/JOSAB.29.002754.

Ben-Aryeh, Y., Y. H. Shih, and M. H. Rubin. 1999. "Optical Wigner Functions for Two-Photon Entangled Wave Packets." *Journal of the Optical Society of America B* 16 (10): 1730–1738.

https://doi.org/10.1364/JOSAB.16.001730.

Byer, R. L. 1975. "Parametric Oscillators and Amplifiers." In *Quantum Electronics: A Treatise*, edited by H. Rabin and C. L. Tang, 587–702. New York: Academic Press.

Carmichael, H. J., G. J. Milburn, and D. F. Walls. 1984. "Squeezing in a Detuned Parametric Amplifier." *Journal of Physics A: Mathematical and General* 17 (2): 469–485. https://doi.org/10.1088/0305-4470/17/2/031.

Dragoman, D. 1997. "I. The Wigner Distribution Function in Optics and Optoelectronics." *Progress in Optics* 37: 1–56. https://doi.org/10.1016/S0079-6638(08)70336-6.

Friberg, S. R., C. K. Hong, and L. Mandel. 1985. "Measurement of Time Delays in the Parametric Production of Photon Pairs." Physical Review Letters 54 (18): 2011–2013. https://doi.org/10.1103/PhysRevLett.54.2011.

Ghosh, R., and L. Mandel. 1987. "Observation of Nonclassical Effects in the Interference of Two Photons." Physical Review Letters 59 (17): 1903–1905. https://doi.org/10.1103/PhysRevLett.59.1903.

Gilmore, R. 1974. *Lie Groups, Lie Algebras, and Some of Their Applications*. New York: Wiley.

Hong, C. K., Z. Y. Ou, and L. Mandel. 1987. "Measurement of Subpicosecond Time Intervals between Two Photons by Interference." Physical Review Letters 59 (18): 2044–2046. https://doi.org/10.1103/PhysRevLett.59.2044.

Huttner, B., S. Serulnik, and Y. Ben-Aryeh. 1990. "Quantum Analysis of Light Propagation in a Parametric Amplifier." *Physical Review A* 42 (9): 5594–5601. https://doi.org/10.1103/PhysRevA.42.5594.

Kim, M. S., F. A. M. de Oliveira, and P. L. Knight. 1989. "Photon Number Distributions for Squeezed Number States and Squeezed Thermal States." *Optics Communications* 72 (2–3): 99–103. https://doi.org/10.1016/0030-4018(89)90263-0.

Kubo, R. 1966. "The Fluctuation-Dissipation Theorem." *Reports on Progress in Physics* 29 (1): 255–284. https://doi.org/10.1088/0034-4885/29/1/306.

Mollow, B. R., and R. J. Glauber. 1967. "Quantum Theory of Parametric Amplification. I." *Physical Review* 160 (5): 1076–1096. https://doi.org/10.1103/PhysRev.160.1076.

Ou, Z. Y., and L. Mandel. 1988. "Observation of Spatial Quantum Beating with Separated Photodetectors." *Physical Review Letters* 61 (1): 54–57. https://doi.org/10.1103/PhysRevLett.61.54.

Pittman, T. B., Y. H. Shih, D. V. Strekalov, and A. V. Sergienko. 1995. "Optical Imaging by Means of Two-Photon Quantum Entanglement." *Physical Review A* 52 (5): R3429–R3432. https://doi.org/10.1103/PhysRevA.52.R3429.

Rubin, M. H., D. N. Klyshko, Y. H. Shih, and A. V. Sergienko. 1994. "Theory of Two-Photon Entanglement in Type-II Optical Parametric Down-Conversion." *Physical Review A* 50 (6): 5122–5133. https://doi.org/10.1103/PhysRevA.50.5122.

Schleich, W., D. F. Walls, and J. A. Wheeler. 1988. "Area of Overlap and Interference in Phase Space versus Wigner Pseudoprobabilities." *Physical Review A* 38 (3): 1177–1189. https://doi.org/10.1103/PhysRevA.38.1177.

Shih, Y. H., and A. V. Sergienko. 1994. "Two-Photon Anti-Correlation in a Hanbury-Brown–Twiss Type Experiment." *Physics Letters A* 186 (1): 29–33. https://doi.org/10.1016/0375-9601(94)90917-2.

Shih, Y. H., and C. O. Alley. 1988. "New Type of Einstein–Podolsky–Rosen–Bohm Experiment." *Physical Review Letters* 61 (26): 2921–2924. https://doi.org/10.1103/PhysRevLett.61.2921.

Toren, M., and Y. Ben-Aryeh. 1994. "The Problem of Propagation in Quantum Optics, with Applications to Amplification, Coupling of EM Modes and Distributed Feedback Lasers." *Quantum Optics: Journal of the European Optical Society Part B* 6 (5): 425–437. https://doi.org/10.1088/0954-8998/6/5/006.

Zahler, M., and Y. Ben-Aryeh. 1991. "Photon Number Distribution of Detuned Squeezed States." *Physical Review A* 43 (12): 6368–6375. https://doi.org/10.1103/PhysRevA.43.6368.

Zahler, M., and Y. Ben-Aryeh. 1992. "Photon Number Distribution of Detuned Two-Mode Vacuum and Excited Squeezed States." Physical Review A 45 (5): 3194–3202. https://doi.org/10.1103/PhysRevA.45.3194

CHAPTER 5

REDUCTION OF QUANTUM NOISE IN MICHELSON INTERFEROMETER

Small changes in the Michelson interferometer's arm-length difference can be observed either by subtle shifts in the interference pattern or through photon statistics measurements. There is a significant interest in developing such interferometers for the detection of gravitational waves, which induce extremely small variations in the interferometer's arm length. Considerable effort has been devoted to overcoming all the technological challenges associated with these sensitive measurements. The first observation of gravitational waves on Earth occurred in September 2015, when two detectors, LIGO's Hanford (Washington) and Livingston (Louisiana), simultaneously recorded the signal (Abbott et al. 2016).

In the typical use of the coherent Michelson interferometer, a coherent state is introduced into one input port while a vacuum state enters the other. The presence of the vacuum state leads to additional quantum fluctuations. Caves (1980; 1981) analyzed the properties of the Michelson interferometer in which a squeezed-vacuum (SV) state replaces the vacuum state entering the interferometer and showed the advantage of using such a scheme.

Following (Caves 1981), the state of the EM field at the entrance to the interferometer can be given by (Scully and Zubairy 1997):

$$|\psi_{in}\rangle = \hat{S}_2(\xi)\hat{D}_1(\alpha)|0\rangle, \tag{5.1}$$

where $\hat{D}_1(\alpha)$ is the displacement operator for the coherent state mode entering input port 1, and $\hat{S}_2(\xi)$ is the squeeze displacement operator in input port 2. These are given by:

$$\hat{D}_1(\alpha)=\exp\left(\alpha\hat{a}_1^{\dagger}-\alpha^{*}\hat{a}_1\right)\ ;\ |\alpha\rangle=\hat{D}_1(\alpha)|0\rangle, \tag{5.2}$$

$$\hat{S}_2(\varsigma)=\exp\left(\frac{1}{2}\varsigma^{*}\hat{a}_2^{\ 2}-\frac{1}{2}\varsigma\hat{a}_2^{\dagger 2}\right)\ ;\ \varsigma=re^{i\theta}, \tag{5.3}$$

where α and ς are complex numbers (the properties of these operators are described in Chapter 1). We assume that the coherent field has an intensity much greater than that of the SV state, although the SV state exhibits large fluctuations. Since the operators $\hat{S}_2(\xi)$ and $\hat{D}_1(\alpha)$ commute, one can interchange their ordering. However, for the present calculations, the order given by Eq. (5.1) is more convenient.

For an interferometer with 50:50 BS (assuming no phase shift in the unitary transformation), we obtain:

$$\begin{pmatrix}\hat{a}_1\\ \hat{a}_2\end{pmatrix}=\begin{pmatrix}\cos\gamma & \sin\gamma\\ -\sin\gamma & \cos\gamma\end{pmatrix}\begin{pmatrix}\hat{b}_1\\ \hat{b}_2\end{pmatrix}, \tag{5.4}$$

where $\hat{b}_1$ and $\hat{b}_2$ are the output operators from the BS. By straightforward calculation, we find the output state $|\psi_{out}\rangle$ from the BS to be:

$$|\psi_{out}\rangle=\exp\left(\hat{A}+\hat{B}\right)D_1(\beta_1)D_2(\beta_2), \tag{5.5}$$

where:

$$\beta_1=\alpha\cos\gamma\ ;\ \beta_2=\alpha\sin\gamma\ ;\ \hat{B}=\sin\gamma\cos\gamma\left(\varsigma\hat{b}_1^{\dagger}\hat{b}_2^{\dagger}-\varsigma^{*}\hat{b}_1\hat{b}_2\right)\ ;$$
$$\hat{A}=\frac{1}{2}\sin^2\gamma\left(\varsigma^{*}b_1^{\ 2}-\varsigma b_1^{\dagger 2}\right)+\frac{1}{2}\cos^2\gamma\left(\varsigma^{*}b_2^{\ 2}-\varsigma b_2^{\dagger 2}\right). \tag{5.6}$$

Notably, the operators $\hat{b}_1^{\dagger}\hat{b}_2^{\dagger}$ and $\hat{b}_1\hat{b}_2$ appearing in $\hat{B}$, produce *entanglement between the two outputs of the interferometer. Such entanglement was not considered in (Caves 1980; Caves 1981), and therefore, the present analysis is different from that of Caves.*

The important special case:

$$\gamma = \pi / 2 + \delta \, , \tag{5.7}$$

where δ is a small phase shift is investigated. For this case, disentanglement methods were developed in Barak and Ben-Aryeh (2008) and Voronov and Weyrauch (2010), through which the output state is given by products of operators in the two separate output ports:

$$\left|\psi_{out}\right\rangle = \hat{D}_1\left[-\alpha\delta\left(1-\cosh r\right)\right]\hat{D}_1\left[-\delta\alpha^{*}e^{i\theta}\sinh r\right]\hat{S}_1\left(\varsigma\right)\hat{D}_1(-\alpha\delta)\hat{D}_2(\alpha)\left|0\right\rangle_1\left|0\right\rangle_2 , \tag{5.8}$$

with $\varsigma = r^{i\theta}$. In the original work (Barak and Ben-Aryeh 2008), a sign error appeared in Eq. (5.8), which affected some of the results and was later treated in Voronov and Weyrauch (2010). We find the interesting result that the output in port 2 is the coherent state $\left|\alpha\right\rangle$ while in output port 1, the operators with index 1 may be combined using the relations provided by Scully and Zubairy (1997) and Voronov and Weyrauch (2010):

$$\begin{gathered}\hat{D}\left(\alpha_2\right)\hat{D}\left(\alpha_1\right) = \hat{D}\left(\alpha_1+\alpha_2\right)\exp\left[\frac{1}{2}\left(\alpha_2\alpha_1^{*}-\alpha_2^{*}\alpha_1\right)\right], \\ \hat{D}\left(\alpha\right)\hat{S}\left(\varsigma\right) = \hat{S}\left(\varsigma\right)\hat{D}\left(\alpha\cosh r+\alpha^{*}e^{i\theta}\sinh r\right) \; ; \; \varsigma = re^{i\theta} .\end{gathered} \tag{5.9}$$

We simplify the analysis by assuming: $\left|\alpha\right\rangle = \left|\alpha\right\rangle^{i\phi} ; \quad \theta - 2\phi = 0$. Then, in the second relation of Eq. (5.9), we can estimate:

$$\begin{aligned}&\alpha\cosh r+\alpha^{*}e^{i\theta}\sinh r \\ &= \left|\alpha\right|e^{i\phi}\left[\cosh r+\exp i\left(\theta-2\phi\right)\sinh r\right] \approx \left|\alpha\right|e^{r} .\end{aligned} \tag{5.10}$$

Here, we ignore the phase term, which is irrelevant to photon statistics. Under these conditions, the second relation in Eq. (5.9) is reduced to:

$$\hat{D}\left(\alpha\right)\hat{S}\left(\varsigma\right) = \hat{S}\left(\varsigma\right)\hat{D}\left(\alpha e^{r}\right). \tag{5.11}$$

By moving the displacement operators in Eq. (5.8) to the right side of $\hat{S}_1(\varsigma)$, and after neglecting phase terms that are not important to photon statistics, we obtain:

$$|\psi_{out}\rangle = \hat{S}_1(\varsigma)D_1(-\alpha\delta e^r)D_2(\alpha)|0\rangle_1|0\rangle_2 . \tag{5.12}$$

This result is equivalent to that obtained by Voronov and Weyrauch (2010) under the above special conditions.

We now apply the above analysis to the detection of gravitational waves, where $\alpha\delta e^r$ is the phase shift produced by the gravitational waves under the specified optimal conditions. The output $\hat{S}_1(\varsigma)D_1(-\alpha\delta e^r)$ in the detection port corresponds to the EM perturbation induced by the gravitational waves. For $\delta = 0$, it represents the "dark output" as it contains only the SV state photons. The average number of photons in the SV state is $\sinh^2 r$, while the *light intensity* in the dark port is enhanced by the use of SV state by the factor e^{2r}. Although one can increase the *gravitational amplitude wave signal* by increasing α, i.e., *increasing the amplitude of the coherent field*, this would also involve an increase in the radiation pressure (to be described later in this chapter). In contrast, enhancing the signal by the squeezing parameter e^r does not result in any change in radiation pressure.

The fact that the optimal condition for detecting gravitational waves is to observe their effect in the dark port is well known. In the above analysis, we have shown the advantage of using squeezed states, where the approach is based on the disentangling of the correlations produced between the two outports.

Vacuum squeezing in gravitational-wave detection observatories has been applied in various studies (see, e.g., Vahlbruch et al. 2008; McKenzie et al. 2002; LIGO Scientific Collaboration 2011) for quantum noise reduction. A comprehensive review on quantum noise in gravitational-wave interferometers was published by Corbitt and Mavalvala (2004), where the prospects for "beating the standard quantum limit" were also explored. This topic is discussed further in the following section.

5.1. The standard quantum limit and the possibility to beat it

"Quantum mechanics places a limit on the accuracy of any measurement of the position of a free mass. In a measurement of duration τ, the minimum possible error in determining z is the *standard quantum limit*:

$$(\Delta z)_{SQL} = \left(\frac{2\hbar\tau}{m}\right)^{1/2}, \tag{5.13}$$

a limit which follows immediately from the Heisenberg uncertainty principle" (Caves 1980):

$$\Delta p \Delta z = \frac{m\Delta z}{\tau}\Delta z = \hbar \rightarrow \left(\frac{\langle \Delta z^2 \rangle}{2}\right)_{SQL} = \frac{\hbar\tau}{m}. \tag{5.14}$$

There is a certain confusion in the literature regarding the concept of "beating the quantum limit". The operators Δz and Δp are considered as "two-quadrature", where one can measure one quadrature to be more accurate at the expense of the noise entering the other quadrature. However, the uncertainty of a two-quadrature state still satisfies the Heisenberg uncertainty principle, as expressed by Caves. This conclusion is valid, however, only when there are no correlations between the two quadratures; otherwise, the state is not considered a "real two-quadrature" system.

Caves also argued that the quantum limit arises from balancing the error due to photon counting statistics against the disturbance of the end-mirrors' positions produced by fluctuating radiation pressure. The laser has a mean power P, so that during a measurement time τ the mean number of photons passing through the interferometer is $N = P\tau / \hbar\omega$. Fluctuations in the laser power intensity produce uncertainty in the number of photons, estimated as $\Delta N = 1/\sqrt{N}$, so that the accuracy related to PC errors is restricted by $\Delta(\delta\varphi) \approx 1/\sqrt{N}$. One should consider that for a coherent state in free space, the number of photons is given by $|\alpha|^2$ within the coherence time, here it is given for the time τ of measurement, shorter than the laser coherence time. Moreover, τ should also be smaller than the period time T of the gravitational wave.

This uncertainty is calculated as $\Delta(\delta\varphi)=(2b\Delta z_{pc}/\lambda)2\pi=(2b\omega/c)(\Delta z_{pc})$, where b is the number of reflections at each mirror, and Δz_{pc} is the difference in the lengths of the two interferometer arms. By equating the two expressions for $\Delta(\delta\varphi)$ we obtain:

$$(2b\omega/c)\left(\Delta z_{pc}\right)\approx 1/\sqrt{N}\rightarrow\left(\Delta z_{pc}\right)\approx\left(c/2b\omega\right)\left(1/\sqrt{N}\right). \tag{5.15}$$

During the measurement time, the radiation pressure induces uncertainty in the difference between the two arm lengths of the interferometer, given by:

$$\left(\Delta z\right)_{rp}\approx\left(\frac{2\hbar\omega b}{c}\right)\left(\frac{\tau}{m}\right)N^{1/2}. \tag{5.16}$$

Here, $N^{1/2}$ is the uncertainty in the number of photons $(2\hbar\omega)/c$, is the momentum transferred to the mirror by a reflected photon b, is the number of bounces at each mirror, and τ is the measurement time. Caves claimed (Caves 1980) that the maximal total error in Δz given by:

$$\Delta z=\left[\left(\Delta z\right)^{2}_{pc}+\left(\Delta z\right)^{2}_{rp}\right]^{1/2}, \tag{5.17}$$

leads to the standard quantum limit given as expressed in Eq. (5.13). However, this relation holds only under the condition that there are no correlations between the two sources of uncertainty.

5.2. Unified model

We simplify the analysis by assuming simple BS transformations for the single-mode annihilation operators, $\hat{a}$ for the coherent state and $\hat{a}_S$ for the squeezed state (as defined in Eq. (1.11)), which are inserted in the two input ports of the BS (Loudon 1981; Assaf and Ben-Aryeh 2002a; Assaf and Ben-Aryeh 2002b) :

$$\hat{a}_1 = \frac{1}{\sqrt{2}}(\hat{a} + \hat{a}_S) \quad ; \quad \hat{a}_2 = \frac{1}{\sqrt{2}}(\hat{a} - \hat{a}_S) \, . \tag{5.18}$$

The annihilation operator for the light that reaches the detector is:

$$\hat{d} = 2^{-1/2}\left[\exp(ik\hat{z}_1)\hat{a}_1 + \exp(ik\hat{z}_2)\hat{a}_2\right], \tag{5.19}$$

where $\hat{z}_1$ and $\hat{z}_2$ are the optical path-length operators for arms 1 and 2, respectively.

In gravitational-wave detection, the mirrors are free to move in directions parallel to the arms of the interferometer. Radiation pressure, thus, causes $\hat{z}_1$ and $\hat{z}_2$ to depend on the intensity of the light at the mirrors. This dependence is expressed as (Loudon 1981):

$$k\hat{z}_1 \rightarrow k\hat{z}_1 + C\hat{a}_1^{\dagger}\hat{a}_1 \quad ; \quad k\hat{z}_2 \rightarrow k\hat{z}_2 + C\hat{a}_2^{\dagger}\hat{a}_2 \, , \tag{5.20}$$

where C is a constant calculated in Assaf and Ben-Aryeh (2002b).

The additional terms, which are proportional to the photon number operators in the two arms, describe the effect of radiation pressure. For any Boson operator, we have:

$$\hat{a}\exp\left(iC\hat{a}^{\dagger}\hat{a}\right) = \exp(iC)\exp\left(iC\hat{a}^{\dagger}\hat{a}\right) . \tag{5.21}$$

The simple transformations from the operators $\hat{a}_1$ and $\hat{a}_2$ to the output operators $\hat{d}$ and $\hat{e}$ take into account the respective arm lengths Z_1 and Z_2, as well as the Kerr-type radiation pressure interactions, and are given in the form:

$$\hat{d} = \frac{1}{\sqrt{2}}\begin{bmatrix} \exp\left(ikZ_1 + iC\hat{a}_1^{\dagger}\hat{a}_1\hat{a}_1\right) \\ +\exp\left(ikZ_2 + iC\hat{a}_2^{\dagger}\hat{a}_2\hat{a}_2\right) \end{bmatrix}, \tag{5.22}$$

$$\hat{e} = \frac{1}{\sqrt{2}}\begin{bmatrix} \exp\left(ikZ_1 + iC\hat{a}_1^{\dagger}\hat{a}_1\hat{a}_1\right) \\ -\exp\left(ikZ_2 + iC\hat{a}_2^{\dagger}\hat{a}_2\hat{a}_2\right) \end{bmatrix} . \tag{5.23}$$

The above unified model conditions were used in our previous works (Assaf and Ben-Aryeh 2002a; Assaf and Ben-Aryeh 2002b), assuming $C \ll 1$, and that the intensity of the coherent field is much larger than the intensity of the squeezed state. However, the squeezed state still exhibits significant quantum fluctuations.

Using the above conditions, a unified model was developed in our previous works (Assaf and Ben-Aryeh 2002a; Assaf and Ben-Aryeh 2002b). We found that a change δZ in the difference between the arm lengths produces a signal differential:

$$\delta\left(\hat{d}^{\dagger}\hat{d}-\hat{e}^{\dagger}\hat{e}\right)=-k\left[\sin k\left(Z_2-Z_1\right)\right]|\alpha|^2\,\delta Z \,. \tag{5.24}$$

One should consider that, for simplicity of notation, we have defined the phase difference in the Michelson interferometer as $k(Z_2-Z_1)$, where this difference should include all additional constant phase shifts that may result from the properties of the mirrors and the BS. Also, Z_2-Z_1 represents the path length difference for one round trip of the photons in the interferometer. When the number of bounces b in the interferometer is larger than one, the constant b can be included in the final expressions. The optimal signal differential is obtained under the condition $\sin k(Z_2-Z_1)=1$, which follows from the fact that the signal is proportional to $\cos k(Z_2-Z_1)$. These conditions apply to a working point in the interferometer where the light intensities in the two outports are equal. Such a condition is fundamentally different from the previous analysis, which assumed a dark output port. We note that the effect of the squeezed state on the signal differential vanishes, as given by Eq. (5.24). However, the squeezed state still has a strong effect on the quantum light intensity fluctuations, which can be observed using the present homodyne detection scheme. After lengthy but straightforward calculations, we obtained the minimal value of the expectation value of the homodyne signal $\langle(\hat{d}^{\dagger}\hat{d}-\hat{e}^{\dagger}\hat{e})^2\rangle_{\min}$ by choosing a special value for $\xi=\theta-2f$:

$$\left\langle\left(\hat{d}^{\dagger}\hat{d}-\hat{e}^{\dagger}\hat{e}\right)^2\right\rangle_{\min}=\exp(-2r)\left(|\alpha|^2+C^2|\alpha|^6\right)\;;\;C \ll 1 \,. \tag{5.25}$$

The first term on the right-hand side of Eq. (5.25) represents the PC fluctuations. By expanding the radiation pressure term to the power of C, we find that the second term, which is proportional to C^2, represents the quantum noise fluctuations caused by the radiation pressure. We get an important result by choosing a certain value for ξ, the use of squeezed states can reduce both PC and radiation pressure fluctuations. Our analysis shows that by choosing $\xi = \pi$, we get:

$$\left\langle \left(\hat{d}^{\dagger}\hat{d} - \hat{e}^{\dagger}\hat{e} \right)^2 \right\rangle_{\min} = |\alpha|^2 \exp(-2r) + |\alpha|^2 C^2 |\alpha|^6 \ ; \ C \gg 1 \quad . \tag{5.26}$$

In this case, however, the noise due to radiation pressure increases. These results are different from those obtained by Caves (1980).

The fact that one can reduce the total noise in Michelson interferometer below the "standard quantum limit" has also been demonstrated in other works (see, e.g., Luis and Sánchez-Soto 1992). Various properties of the Michelson interferometer in relation to gravitational wave detection are published in a book devoted to this topic (Saulson 1994).

References

Abbott, B. P., R. Abbott, T. D. Abbott, M. R. Abernathy, F. Acernese, K. Ackley, C. Adams, et al. 2016. "Observation of Gravitational Waves from a Binary Black Hole Merger." *Physical Review Letters* 116 (6): 061102. https://doi.org/10.1103/PhysRevLett.116.061102.

Assaf, O., and Y. Ben-Aryeh. 2002A. "Quantum Mechanical Noise in Coherent State and Squeezed State Michelson Interferometer." *Journal of Optics B: Quantum and Semiclassical Optics* 4 (1): 49–54. https://doi.org/10.1088/1464-4266/4/1/308.

Assaf, O., and Y. Ben-Aryeh. 2002B. "Reduction of Quantum Noise in the Michelson Interferometer by Use of Squeezed Vacuum State." *Journal of the Optical Society of America* B 19 (11): 2716–2721. https://doi.org/10.1364/JOSAB.19.002716.

Barak, R., and Y. Ben-Aryeh. 2008. "Photon Statistics and Entanglement in Coherent-Squeezed Linear Mach-Zehnder and Michelson Interferometer." *Journal of the Optical Society of America B* 25 (3): 361–370. https://doi.org/10.1364/JOSAB.25.000361.

Caves, C. M. 1980. "Quantum-Mechanical Radiation-Pressure Fluctuations in an Interferometer." *Physical Review Letters* 43 (1): 75–79.

https://doi.org/10.1103/PhysRevLett.45.75.
Caves, C. M. 1981. "Quantum Mechanical Noise in an Interferometer." *Physical Review A* 23 (4): 1693–1708. https://doi.org/10.1103/PhysRevD.23.1693.
Corbitt, T., and N. Mavalvala. 2004. "Review: Quantum Noise in Gravitational-Wave Interferometers." *Journal of Optics B: Quantum and Semiclassical Optics* 6 (8): S675–S683. https://doi.org/10.1088/1464-4266/6/8/008.
LIGO Scientific Collaboration. 2011. "A Gravitational Wave Observatory Operating Beyond the Quantum Shot-Noise Limit." *Nature Physics* 7 (12): 962–965. https://doi.org/10.1038/nphys2083.
Loudon, R. 1981. "Quantum Limit on the Michelson Interferometer Used for Gravitational-Wave Detection." *Physical Review Letters* 47 (11): 815–818. https://doi.org/10.1103/PhysRevLett.47.815.
Luis, A., and L. L. Sánchez-Soto. 1992. "Breaking the Standard Quantum Limit for Interferometric Measurements." *Optics Communications* 89 (1–2): 140–144. https://doi.org/10.1016/0030-4018(92)90147-J.
McKenzie, K., D. A. Shaddock, D. E. McClelland, B. C. Buchler, and P. K. Lam. 2002. "Experimental Demonstration of a Squeezing-Enhanced Power-Recycled Michelson Interferometer for Gravitational Wave Detection." *Physical Review Letters* 88 (23): 231102. https://doi.org/10.1103/PhysRevLett.88.231102.
Saulson, P. R. 1994. Fundamentals of Interferometric Gravitational Wave Detectors. *Singapore: World Scientific.*
Scully, M. O., and M. S. Zubairy. 1997. *Quantum Optics*. Cambridge: Cambridge University Press.
Vahlbruch, H., M. Mehmet, S. Chelkowski, B. Hage, A. Franzen, N. Lastzka, S. Gobler, K. Danzmann, and R. Schnabel. 2008. "Observation of Squeezed Light with 10-dB Quantum Noise Reduction." *Physical Review Letters* 100 (3): 033602. https://doi.org/10.1103/PhysRevLett.100.033602.
Voronov, V. G., and M. Weyrauch. 2010. "Quantum Noise in Optical Interferometers." *Physical Review A* 81 (5): 053816. https://doi.org/10.1103/PhysRevA.81.053816.

CHAPTER 6

GENERAL REPRESENTATIONS IN QUANTUM OPTICS

The density operator may be represented in the diagonal coherent state representation known as the *P*-representation (Glauber 1963; Sudarshan 1963), according to

$$\hat{\rho} = \int P(\alpha)|\alpha\rangle\langle\alpha| d^2\alpha \quad , \tag{6.1}$$

where $P(\alpha)$ is a quasi-probability distribution function (which need not be positive definite), and the integration is made over the complex plane. The one mode $|\alpha\rangle$ state is the eigenstate of the annihilation operator $\hat{a}$, which can be obtained from the vacuum state $|0\rangle$ via the unitary displacement operator $\hat{D}(\alpha)$:

$$\hat{a}|\alpha\rangle = \alpha|\alpha\rangle \quad ; \quad |\alpha\rangle = D(\alpha)|0\rangle = \exp\left[\alpha\hat{a}^{\dagger} - \alpha^{*}\hat{a}^{\dagger}\right]|0\rangle \quad . \tag{6.2}$$

The coherent state $|\alpha\rangle$ has a photon number distribution given by:

$$\langle n|\alpha\rangle = \alpha^{n}(n!)^{-1/2}\exp\left[-|\alpha|^{2}/2\right] \quad . \tag{6.3}$$

Coherent states are not orthogonal and form an overcomplete basis, explicitly:

$$\langle\beta|\alpha\rangle| = \exp\left[-\frac{1}{2}\left(|\alpha|^{2} + |\beta|^{2}\right) + \beta^{*}\alpha\right] \quad . \tag{6.4}$$

Consequently, it is possible to express the identity operator as:

$$I = \frac{1}{\pi}\int d^2\alpha |\alpha\rangle\langle\alpha| \,, \tag{6.5}$$

where the integration is performed over the complex plane. Any operator $\hat{A}(\hat{a},\hat{a}^\dagger)$ which is expanded in a *normally ordered power series*:

$$\hat{A} = \sum_{n,m=0}^{\infty} c_{n,m}\left(\hat{a}^\dagger\right)^n \hat{a}^m \,, \tag{6.6}$$

fulfills the relation:

$$Tr\left(\hat{\rho}\hat{A}\right) = \int d^2\alpha P(\alpha)\langle\alpha|A|\alpha\rangle = \int d^2\alpha P(\alpha) A_N\left(\alpha,\alpha^*\right);$$
$$A_N\left(\alpha,\alpha^*\right) = \sum_{n,m=0}^{\infty} c_{nn}\left(\alpha^*\right)^n \alpha^m . \tag{6.7}$$

We can use the *P*-representation to calculate the function $g^{(2)}(0)-1$, as developed and explained in Eqs. (2.8-2.13) of Chapter 2:

$$g^{(2)}(0)-1 = \frac{\int P(\alpha)\left[|\alpha|^2 - \left\langle|\alpha|^2\right\rangle\right]^2 d^2\alpha}{\left[\int P(\alpha)|\alpha|^2 d^2\alpha\right]^2} \,, \tag{6.8}$$

where $\left\langle|\alpha|^2\right\rangle = \int P(\alpha) d^2\alpha$, and for non-classical states, one obtains $g^{(2)}(0)-1<0$. However, the $P(\alpha)$ distribution function has singularity problems for certain types of density matrices.

The properties of quantum states can be related to the $Q(\alpha)$ distribution function, defined as:

$$Q(\alpha) = \pi^{-1}\langle\alpha|\hat{\rho}|\alpha\rangle \,, \tag{6.9}$$

where $\hat{\rho}$ is the density matrix and $|\alpha\rangle$ represents the overcomplete set of coherent states. Any operator $\hat{A}(\hat{a},\hat{a}^\dagger)$ which is expanded in an *anti-normally (AN) ordered series*, takes the form:

$$\hat{A} = \sum_{n,m=0}^{\infty} d_{nm} \hat{a}^m \left(\hat{a}^\dagger\right)^n \quad , \tag{6.10}$$

and fulfills the relation (Hillery et al. 1984):

$$Tr\left(\hat{\rho}\hat{A}\right) = \int d^2\alpha A_{AN}\left(\alpha,\alpha^*\right) Q(\alpha) \; ; \tag{6.11}$$

$$A_{AN}\left(\alpha,\alpha^*\right) = \sum_{n,m=0}^{\infty} d_{nm} \alpha^m \left(\alpha^*\right)^n \; . \tag{6.12}$$

For a coherent state $|\alpha'\rangle$ we obtain the Gaussian distribution:

$$Q(\alpha) = \pi^{-1} \exp\left[-|\alpha - \alpha'|^2\right] . \tag{6.13}$$

For SS the distribution remains Gaussian but becomes non-isotropic (Yuen 1976).

We can introduce symmetry ordering as a product of m annihilation operators and n creation operators, which can be ordered in $(n+m)!/n!m!$ ways. The symmetrically ordered (S) product of these operators, denoted by $\{(a^\dagger)^n \hat{a}^m\}$, is the average of all these differently ordered products. For example: $\{\hat{a}^\dagger \hat{a}^2\} = (\hat{a}^\dagger \hat{a}^2 + \hat{a}\hat{a}^\dagger \hat{a} + \hat{a}^2 \hat{a}^\dagger)/3$. The operator-function correspondence is similar to the previous ones. Specifically, any operator $\hat{A}_S(\hat{a},\hat{a}^\dagger)$ can be expanded in a symmetrically ordered (S) power series:

$$\hat{A}_S\left(\hat{a},\hat{a}^\dagger\right) = \sum_{n,m=0}^{\infty} b_{n,m} \left\{\left(\hat{a}^\dagger\right)^n \hat{a}^m\right\} . \tag{6.14}$$

The function corresponding to this operator is then given by:

$$\hat{A}_S\left(\alpha,\alpha^*\right) = \sum_{n,m=0}^{\infty} b_{n,m} \left\{\left(\alpha^*\right)^n \alpha^m\right\} . \tag{6.15}$$

From this, we obtain:

$$Tr\left(\hat{A}_S\rho\right)=\frac{1}{\pi}\int d^2\alpha A_S\left(\alpha,\alpha^{*}\right)W\left(\alpha\right), \tag{6.16}$$

where $W\left(\alpha\right)$ is the Wigner quasi-probability distribution, which can be derived as (Hillery et al. 1984):

$$W(\alpha)=2\int d^2\beta\exp\left[-2\left|\alpha-\beta\right|^2\right]P(\beta). \tag{6.17}$$

Let $\chi_N\left(\eta^{*},\eta\right)$ be the normally ordered characteristic function:

$$\chi_N\left(\eta^{*},\eta\right)=Tr\left[\rho\exp\left(\eta\hat{a}^{\dagger}\right)\exp\left(-\eta^{*}\hat{a}\right)\right]. \tag{6.18}$$

Then, the quasi-probability distribution $P(\alpha,\alpha^{*})$ is given by the Fourier transform:

$$P\left(\alpha,\alpha^{*}\right)=\int\exp\left[\eta^{*}\alpha-\alpha^{*}\eta\right]\chi_N\left(\eta^{*},\eta\right)d^2\eta/\pi^2. \tag{6.19}$$

Other characteristic functions are:

$$\chi_{AN}\left(\eta^{*},\eta\right)=Tr\left[\rho\exp\left(-\eta^{*}\hat{a}\right)\exp\left(\eta\hat{a}^{\dagger}\right)\right];$$
$$\chi_W\left(\eta^{*},\eta\right)=Tr\left[\rho\exp\exp\left(-\eta^{*}\hat{a}+\eta\hat{a}^{\dagger}\right)\right], \tag{6.20}$$

where their transforms are, respectively, the Wigner and the $Q(\alpha)$ quasi-distribution functions:

$$W\left(\alpha,\alpha^{*}\right)=\int\chi_W\left(\eta^{*},\eta\right)\exp\left[\eta^{*}\alpha-\alpha^{*}\eta\right]d^2\eta/\pi^2\ ;$$
$$Q\left(\alpha,\alpha^{*}\right)=\int\chi_{AN}\left(\eta^{*},\eta\right)\exp\left[\eta^{*}\alpha-\alpha^{*}\eta\right]d^2\eta/\pi^2. \tag{6.21}$$

Other distribution functions have been proposed, but they have found limited practical use (Hillery et al. 1984). A more extensive analysis, however, has been made using the generalized *P*-representations developed by Drummond, Gardiner, and Walls (1981).

6.1. Determination of quasi-probability distributions in terms of distributions for the rotated quadrature phases

Instead of using the three different quasi-probability distributions discussed above, we can use the s-parametrized quasi-probability distributions introduced by Cahill and Glauber (1969). These distributions are given by the Fourier transforms:

$$W(\alpha,s)=\frac{1}{\pi^2}\int \tilde{W}(\xi,s)\exp\left(\alpha\xi^*-\alpha^*\xi\right)d^2\xi \, , \qquad (6.22)$$

where $\tilde{W}(\xi,s)$ are the characteristic functions, given by:

$$\tilde{W}(\xi,s)=Tr\left\{\exp\left(\xi\hat{a}^\dagger-\xi^*\hat{a}+s|\xi|^2/2\right)\hat{\rho}\right\}, \qquad (6.23)$$

and $\hat{\rho}$ is the density matrix. Here for $s=1$ we get the Glauber-Sudarshan P-function, for $s=0$ we get the Wigner function, and for $s=-1$ we get the $Q=\langle\alpha|\hat{\rho}|\alpha\rangle/\pi$ function.

For a light mode with an annihilation operator $\hat{a}$, the operators $\hat{x}$ and $\hat{p}$ are defined by:

$$\hat{x}=\frac{\left(\hat{a}+\hat{a}^\dagger\right)}{\sqrt{2}}; \quad \hat{p}=\frac{i\left(\hat{a}-\hat{a}^\dagger\right)}{\sqrt{2}}. \qquad (6.24)$$

The quadrature amplitudes are represented by the operators:

$$\hat{x}_\theta=\hat{x}\cos\theta-\hat{p}\sin\theta \; ; \; \hat{p}_\theta=\hat{x}\sin\theta+\hat{p}\cos\theta . \qquad (6.25)$$

We refer to the operator $\hat{x}_\theta$ as the rotated quadrature phase. Complete information for calculating any expectation value of the rotated quadrature phase $\hat{x}_\theta$ is given by the probability distribution $w(x,\theta)$. A one-to-one correspondence exists between the quasi-probability distributions and the distribution $w(x,\theta)$ as developed in Vogel and Risken (1989) and Smith et al. (1993). We present here this correspondence for the Wigner function, i.e., for $s = 0$, which has been directly applied in experimental measurements (e.g., Yuen and Chan 1983).

By changing the complex parameter α of Eq. (6.22) to the quadrature parameters, one gets for the Wigner function, after some derivations, the simple expression (Vogel and Risken 1989, Smith et al. 1993):

$$w(x,\theta) = \int_{-\infty}^{\infty} W(x_\theta \cos\theta - p_\theta \sin\theta, x_\theta \sin\theta + p_\theta \cos\theta) dp_\theta \quad . \tag{6.26}$$

where $w(x,\theta)$ can be obtained by homodyne detection (Schomaker 1984; Wigner 1932). The inverse of this relation was also developed as (Vogel and Risken 1989, Smith et al. 1993):

$$W(\alpha_r, \alpha_i, s=0) =$$
$$\int_{-\infty}^{\infty}\int_{-\infty}^{\infty}\int_{0}^{\pi} w(x,\theta) \exp\left[i\eta\left(x - \alpha_r \cos\theta - \alpha_i \sin\theta\right)\right] |\eta|\, dx d\eta d\theta. \tag{6.27}$$

Where for $W(\alpha, s)$ is given by Eqs. (6.22-6.23). We considered the case for the Wigner function by setting $s = 0$, and we separated α into its real and imaginary parts. The explicit use of homodyne detection for deriving $w(x,\theta)$ is quite complicated; however, it can be simplified under certain special conditions (Schomaker 1984; Wigner 1932). As is well known, the Wigner function, for a single degree of freedom, is given by (Lutterbach and Davidovich 1997):

$$W(x.p) = \frac{1}{\pi} \int_{-\infty}^{\infty} \langle x + x' | \hat{\rho} | x - x' \rangle \exp(-2ipx) dx' \quad . \tag{6.28}$$

The present methods, with some approximations, were used in (Yuen 1983) for measuring Wigner distributions for SV and vacuum states.

6.2. Measurement of Wigner distributions corresponding to the EM field in cavities and ion traps

Lutterbach and Davidovich (1997) proposed a method that yields the Wigner function W at a given point α in phase space. Their method is based on a simple expression of $W(\alpha)$, at a point α in phase space, for a field with a density matrix $\hat{\rho}$ (Cahill and Glauber 1969):

$$W(\alpha)=2Tr\left[\hat{D}(-\alpha)\right]\hat{\rho}\hat{D}(\alpha)\hat{P}], \tag{6.29}$$

where $\hat{P}=\exp(i\pi\hat{a}^{\dagger}\hat{a})$ is the field parity operator (its action on Fock state $|n\rangle$ is $\hat{P}|n\rangle=(-1)^{n}|n\rangle$) and $D(\alpha)=\exp[\alpha\hat{a}^{\dagger}-\alpha^{*}\hat{a}^{\dagger}]$ is the displacement operator. With this normalization, $-2\leq W\leq 2$, and $\int W(\alpha)d^{2}\alpha=\pi$. For an n-photon Fock state, $W(0)=2(-1)^{n}$. The method was applied to determine the vibrational state of a trapped ion. In this scheme, the measurement is implemented by atomic interferometry, where an atomic superposition state, coupled to the field, experiences a phase shift of π each time the number of photons is increased by 1. This method was also used for the measurement of the Wigner function at the origin of phase space for a single photon field. The result, a negative value of the Wigner function (Nogues et al. 1999), clearly exhibits the nonclassical nature of the state. The experiment is based on the absorption-free detection of the microwave field in a superconductive cavity (Banaszek et al. 1999).

A direct measurement of the Wigner function of a light mode was reported by Royer (1977). Unlike optical homodyne tomography, this technique is based on photon counting. It was later asserted that the Wigner function at a given phase-space point is a well-defined quantum observable (Leibfried et al. 1996). In this method, the amplitude and phase of a probe field define the point in the phase space at which the Wigner function is measured. The phase space is then scanned by systematically changing the parameters of the probe field. The method is implemented in a Mach-Zehnder interferometer, with the beams in the two arms of the interferometer serving as the signal and the probe fields. Using this configuration, Wigner functions of the vacuum, coherent, and diffused coherent states were successfully measured (Royer 1977).

Density matrices and Wigner functions for various quantum states of motion of a harmonically bound He-ion were reconstructed (Gottfried 1966). The methods involved applying coherent displacements of different amplitudes and phases, followed by measurements of the number state populations. By using reconstruction schemes, both the density matrix and the Wigner function were measured in the number state basis. Notably, negative Wigner functions were found in position-momentum space (Gottfried 1966).

6.3. SU(1,1) coherent and intelligent states

The Heisenberg uncertainty relation can be generalized for any pair of observables $\hat{A}$ and $\hat{B}$, given by:

$$(\Delta A)^2 (\Delta B)^2 \geq \frac{1}{4}\langle\psi|\left[\hat{A},\hat{B}|\psi\rangle\right]^2, \tag{6.30}$$

where the variance $(\Delta A)^2 = \langle\hat{A}^2\rangle - \langle\hat{A}\rangle^2$, $(\Delta B)^2$ is defined similarly. The expectation values are taken with respect to the state of the system $|\psi\rangle$. States for which equality holds in Eq. (6.30) are called intelligent states (IS). In the general case $[\hat{A},\hat{B}] = iC$ and the IS are determined from the eigenvalue equation (Aragone, Chalbaud, and Salamo 1976):

$$\left(\hat{A} + i\gamma\hat{B}\right)|\lambda\rangle_{AB} = \lambda|\lambda\rangle_{AB}\ ;\ \left[\hat{A},\hat{B}\right] = iC, \tag{6.31}$$

where λ is a complex eigenvalue,

$$\lambda = \left\langle\hat{A}\right\rangle + i\gamma\left\langle\hat{B}\right\rangle, \tag{6.32}$$

and λ is a real parameter given by:

$$\gamma = \frac{\left|\left\langle\hat{C}\right\rangle\right|}{2(\Delta B)^2} = \frac{2(\Delta A)^2}{\left\langle\hat{C}\right\rangle}. \tag{6.33}$$

The properties of intelligent states were analyzed for the SU(2) (Ruschin and Ben-Aryeh 1976; Brif and Ben-Aryeh 1994a) and SU(1,1) generators (Wódkiewicz and Eberly 1985). The concept of squeezing is related to SU (2) and SU(1,1) IS (Hillery 1987; Perelomov 1972). For $\gamma > 1$, IS are squeezed in $\hat{B}$; for $\gamma < 1$, IS are squeezed in $\hat{A}$.

Unlike the special case of the Heisenberg-Weyl Lie group, the IS states and coherent states are different for the SU(2) and SU(1,1) Lie groups. The Lie algebra corresponding to the group SU(1,1) has 3 generators: $\hat{K}_1$, $\hat{K}_2$, and $\hat{K}_3$ satisfying the commutation relations:

$$\left[\hat{K}_1,\hat{K}_2\right]=-i\hat{K}_3 \quad ; \quad \left[\hat{K}_2,\hat{K}_3\right]=i\hat{K}_1 \quad ; \quad \left[\hat{K}_3,\hat{K}_1\right]=i\hat{K}_2 . \tag{6.34}$$

The Casimir operator is given by:

$$\hat{Q}=\hat{K}_3^{\,2}-\hat{K}_1^{\,2}-\hat{K}_2^{\,2}, \tag{6.35}$$

for any irreducible representation. The identity operator multiplied by the number $k(k-1)$:

$$\hat{Q}=k(k-1)\hat{I}, \tag{6.36}$$

where a representation of SU(1,1) is determined by the single number k.

We discuss the case of the discrete series where the complete orthonormal basis spans the corresponding state space:

$$\begin{aligned}&\left|k,m\right\rangle \; (m=0,1,\cdots,\infty);\\&\left\langle km\middle|k,m'\right\rangle=\delta_{mm'} \quad ; \quad \sum_{m=0}^{\infty}\left|k,m\right\rangle\left\langle k,m\right|=1.\end{aligned} \tag{6.37}$$

These states satisfy the relations:

$$\begin{aligned}&K_3\left|k,m\right\rangle=(k+m)\left|k,m\right\rangle \quad ; \quad K_\pm=K_1\pm iK_2;\\&K_+\left|k,m\right\rangle=\left[\left(m+1(2k+m)\right)\right]^{1/2}\left|k,m+1\right\rangle;\\&K_-\left|k,m\right\rangle=\left[m(2k+m-1)\right]^{1/2}\left|k,m-1\right\rangle.\end{aligned} \tag{6.38}$$

The SU(1,1) discrete series CS (Gerry 1985) are specified by pseudo-Euclidean unit vectors of the form:

$$n=\cosh\tau,\sinh\tau\cos\varphi,\sinh\tau\sin\varphi . \tag{6.39}$$

The CS $|k,\varsigma\rangle$ are obtained by applying a unitary operator $\hat{D}(\xi)$ to the extreme state $|k,m=0\rangle$:

$$|k,\varsigma\rangle=\exp\left(\xi\hat{K}_{+}-\xi^{*}\hat{K}_{-}\right)|k,0\rangle=\left(1-|\varsigma|^{2}\right)^{k}\exp\left(\varsigma\hat{K}_{+}\right)|k,0\rangle\ ;$$
$$\xi=-\frac{\tau}{2}e^{-i\varphi}\ ;\ \varsigma=-\tanh\frac{\tau}{2}e^{-i\varphi}\ ;\ |\varsigma|<1. \tag{6.40}$$

Expanding the exponential and using Eq. (6.38), one obtains the decomposition of the CS over the orthonormal basis:

$$|k,\varsigma\rangle=\left(1-|\varsigma|^{2}\right)^{k}\sum_{m=0}^{\infty}\left[\frac{\Gamma(m+2k)}{m!\Gamma(2k)}\right]\varsigma^{m}|k,m\rangle \tag{6.41}$$

The dynamics of the SU(1,1) CS have been studied by Gerry (1985), who derived the most general Hamiltonian that preserves the Perelomov SU(1,1) coherent states. The Evolution of quantum systems driven by Hamiltonian, which includes linear combinations of the SU(1,1) generators, has been analyzed by using algebraic operational ordering methods (Dattoli, Gallardo, and Torre 1988).

The condition $|\varsigma|<1$ *implies that the 'phase space' of the SU(1,1) CS is confined to the interior of the unit disk in the complex plane.* These coherent states are normalized but not orthogonal, as expressed by:

$$\langle k,\varsigma_1|k,\varsigma_2\rangle=\left(1-|\varsigma_1|^{2}\right)^{k}\left(1-|\varsigma_2|^{2}\right)^{k}\left(1-\varsigma_1^{*}\varsigma_2\right)^{-2k} \tag{6.42}$$

The identity resolution of the coherent states is:

$$\int d\mu(\varsigma)|k,\varsigma\rangle\langle k,\varsigma|=1\ ;\ d\mu(\varsigma)=\frac{2k-1}{\pi}\frac{d^{2}\varsigma}{\left(1-|\varsigma|^{2}\right)^{2}} \tag{6.43}$$

Although the SU(1,1), IS, which are minimum uncertainty states, possess nice physical properties, they are usually realizable only by non-Hermitian Hamiltonians. There are, however, special cases in which IS are coinciding with the SU(1,1) coherent states (Wódkiewicz and Eberly

1985; Brif and Ben-Aryeh 1994b). We considered a SU(1,1) interferometer employing four-wave mixers, fed with two-mode states, which are both coherent and intelligent states of the SU(1,1) Lie group. The phase sensitivity of the interferometer can be improved by using input states with a large photon difference between the modes. An extensive analysis of the use of SU(2) and SU(1,1) interferometers was presented by Yurke, McCall, and Klauder (1986). A new type of SU(1,1) interferometer, which employs parametric amplifiers as the wave splitting and mixing elements, has been recently developed (Ou and Li 2020).

The evolution operator of the Wigner function for the most general quadratic Hamiltonian with time-dependent coefficients was obtained. Utilizing group theoretical methods, the time development of the quantum state was solved, enabling a detailed description of its nonclassical properties (Ben-Aryeh and Zoubi 1996; Zoubi and Ben-Aryeh 1998).

Generalized coherent states, which are the eigenstates of the lowering and raising operators of non-compact groups, were developed by Barut and Girardello (Barut and Girardello 1971). A sub-coherent *P*-representation for nonclassical photon states was later developed by Brif and Ben-Aryeh (1994c). These states exhibit sub-Poissonian photon statistics. By using these states, the diagonal *P*-representation of the density operator was constructed, and it was shown to be well behaved even for non-classical photon states.

References

Aragone, C., E. Chalbaud, and S. Salamo. 1976. "On Intelligent Spin States." *Journal of Mathematical Physics* 17: 1963–1966. https://doi.org/10.1063/1.522835

Banaszek, K., C. Radzewicz, K. Wódkiewicz, and J. S. Krasinski. 1999. "Direct Measurement of the Wigner Function by Photon Counting." *Physical Review A* 60: 674–677. https://doi.org/10.1103/PhysRevA.60.674

Barut, A. O., and L. Girardello. 1971. "New 'Coherent' States Associated with Non-Compact Groups." Communications in Mathematical Physics 21: 41-55. https://doi.org/10.1007/BF01646483

Ben-Aryeh, Y., and H. Zoubi. 1996. "The Time Development Operators for Wigner Functions of Harmonic Oscillators with Quadratic Hamiltonians." Quantum and Semiclassical Optics 8: 1097-1105. https://doi.org/10.1088/1355-5111/8/6/001

Brif, C., and Y. Ben-Aryeh. 1994a. "Sub Coherent p-Representation for a Non-Classical Photons States." Quantum Optics 6: 391-399. https://doi.org/10.1088/0954-8998/6/5/001

Brif, C., and Y. Ben-Aryeh. 1994b. "Improvement of Accuracy in SU(1,1) Interferometer." *Journal of Physics A: Mathematical and General* 27: 8185–8196. https://doi.org/10.1088/0305-4470/27/24/025

Brif, C., and Y. Ben-Aryeh. 1994c. "SU(1,1) Intelligent States: Analytical Representations in the Unit Disk." *Journal of Physics A: Mathematical and General* 27: 8185–8195. https://doi.org/10.1088/0305-4470/27/24/025

Cahill, K. E., and R. J. Glauber. 1969. "Density Operators and Quasiprobability Distributions I & II." *Physical Review* 177: 1857–1881, 1882–1902. https://doi.org/10.1103/PhysRev.177.1857

Dattoli, G., J. C. Gallardo, and A. Torre. 1988. "An Algebraic View to the Optical Ordering and Its Applications to Optics." *Rivista del Nuovo Cimento* 11: 1–79. https://doi.org/10.1007/BF02724503

Drummond, P. D., C. W. Gardiner, and D. F. Walls. 1981. "Quasi Probability Methods for Nonlinear Chemical and Optical Systems." *Physical Review A* 24: 914–936. https://doi.org/10.1103/PhysRevA.24.914

Gerry, C. C. 1985. "Dynamics of SU(1,1) Coherent States." *Physical Review A* 31: 2721–2725. https://doi.org/10.1103/PhysRevA.31.2721

Glauber, R. J. 1963. *The Quantum Theory of Optical Coherence. Physical Review* 130: 2529–2539. https://doi.org/10.1103/PhysRev.130.2529

Gottfried, K. 1966. *Quantum Mechanics*. New York: Benjamin.

Hillery, M. 1987. "Amplitude Squeezing of the Electromagnetic Field." *Physical Review A* 36: 3796–3802. https://doi.org/10.1103/PhysRevA.36.3796

Hillery, M., R. F. O'Connell, M. O. Scully, and E. P. Wigner. 1984. "Distribution Functions in Physics: Fundamentals." *Physics Reports* 106: 121–167. https://doi.org/10.1016/0370-1573(84)90160-1

Leibfried, D., D. M. Meekhof, B. E. King, C. Monroe, W. M. Itano, and D. J. Wineland. 1996. "Experimental Determination of the Motional Quantum State of a Trapped Ion." *Physical Review Letters* 77: 4281–4285. https://doi.org/10.1103/PhysRevLett.77.4281

Lutterbach, L. G., and L. Davidovich. 1997. "Method for Direct Measurement of the Wigner Function in Cavity QED and the Ion Trap." *Physical Review Letters* 78: 2547–2550. https://doi.org/10.1103/PhysRevLett.78.2547

Nogues, G., A. Rauschenbeutel, S. Osnaghi, M. Brune, J. M. Raimond, and S. Haroche. 1999. "Seeing a Single Photon Without Destroying It." *Nature* 400: 239–242. https://doi.org/10.1038/22275

Nogues, G., A. Rauschenbeutel, S. Osnaghi, P. Bertet, M. Brune, J. M. Raimond, S. Haroche, L. G. Lutterbach, and L. Davidovich. 2000. "Measurement of a Negative Value for the Wigner Function of Radiation." *Physical Review A* 62: 054101. https://doi.org/10.1103/PhysRevA.62.054101

Ou, Z. Y., and X. Li. 2020. "Quantum SU(1,1) Interferometers: Basic Principles and Applications." *APL Photonics* 5: 080902. https://doi.org/10.1063/5.0004873

Perelomov, A. M. 1972. "Coherent States for Arbitrary Lie Group." *Communications in Mathematical Physics* 26: 222–236. https://doi.org/10.1007/BF01645091

Royer, A. 1977. "Wigner Function as the Expectation Value of the Parity Operator." *Physical Review A* 13: 449–453. https://doi.org/10.1103/PhysRevA.15.449

Ruschin, S., and Y. Ben-Aryeh. 1976. "Minimum Uncertainty States for Angular Momentum Operators." *Physics Letters A* 58: 207–210. https://doi.org/10.1016/0375-9601(76)90072-4

Schomaker, R. L. 1984. "Noise in Homodyne Detection." *Optics Letters* 9: 189–191. https://doi.org/10.1364/OL.9.000189

Smith, D. T., M. Beck, M. G. Raymer, and A. Faridani. 1993. "Measurement of Wigner Distribution and the Density Matrix of a Light Mode Using Optical Homodyne Tomography: Application to Squeezed States and the Vacuum." *Physical Review Letters* 70: 1244–1247. https://doi.org/10.1103/PhysRevLett.70.1244

Sudarshan, E. C. G. 1963. "Equivalence of Semiclassical and Quantum Mechanical Description of Statistical Light Beams." *Physical Review Letters* 10: 277–279. https://doi.org/10.1103/PhysRevLett.10.277

Vogel, K., and H. Risken. 1989. "Determination of Quasiprobability Distributions in Terms of Probability Distributions for the Rotated Quadrature Phase." *Physical Review A* 40: 2847–2849. https://doi.org/10.1103/PhysRevA.40.2847

Wigner, E. 1932. "On the Quantum Correction for Thermodynamic Equilibrium." *Physical Review* 40: 749–759. https://doi.org/10.1103/PhysRev.40.749

Wódkiewicz, K., and J. H. Eberly. 1985. "Coherent States, Squeezed Fluctuations and the SU(2) and SU(1,1) Groups in Quantum Optics Applications." *Journal of the Optical Society of America B* 2: 458–466. https://doi.org/10.1364/JOSAB.2.000458

Yuen, H. P. 1976. “Two-Photon Coherent States of the Radiation Field.” *Physical Review A* 13: 2226–2243. https://doi.org/10.1103/PhysRevA.13.2226

Yuen, H. P., and V. W. S. Chan. 1983. “Noise in Homodyne and Heterodyne Detection.” *Optics Letters* 8: 177–179. https://doi.org/10.1364/OL.8.000177

Yurke, B., S. L. McCall, and J. R. Klauder. 1986. “SU(2) and SU(1,1) Interferometers.” *Physical Review A* 33: 4033–4054. https://doi.org/10.1103/PhysRevA.33.4033

Zoubi, H., and Y. Ben-Aryeh. 1998. “The Evolution of Harmonic Oscillator Wigner Function Described by Use of Group Representations.” Quantum and Semiclassical Optics 10: 447-454. https://doi.org/10.1088/1355-5111/10/2/016

CHAPTER 7

PHASE OPERATORS OF ELECTROMAGNETIC FIELDS AND PHASE DISTRIBUTIONS

There are several books addressing the foundations of quantum optics, each emphasizing different aspects of this field (e.g., Gerry and Knight 2004; Schleich 2015). One of the problems in quantum optics, still lacking a definite answer, is the 'quantum phase' problem: What exactly is its definition, and how is it related to measurements? Or, more generally, how does it relate to the density matrix, which includes all the information about the quantum state? In this chapter, we explore a few approaches to this fundamental question.

To provide a quantum description of optical phase, Susskind and Glogower (SG) (1964) defined a set of raising and lowering operators within the Hilbert space of the harmonic oscillator:

$$\exp\left(-i\hat{\phi}\right) \equiv \hat{E}_+ = \sum_{n=0}^{\infty} |n+1\rangle\langle n| \ ; \ \exp\left(i\hat{\phi}\right) \equiv \hat{E}_- = \sum_{n=0}^{\infty} |n\rangle\langle n+1| \quad . \tag{7.1}$$

Here, $|n\rangle$ $(n = 0,1,2,\cdots,\infty)$ are the number states. However, since the number states are bounded from below by the vacuum state, the SG operators are not unitary:

$$\hat{E}_-\hat{E}_+ = 1 \quad ; \quad \hat{E}_+\hat{E}_- = 1 - |0\rangle\langle 0| \, . \tag{7.2}$$

This non-unitarity leads to difficulties in the quantum description of the optical phase (Susskind and Glogower 1964, Carruthers and Nieto 1968). Several approaches have been proposed to address this issue. One method involves using a finite-dimensional Hilbert space (Barnett and Pegg 2007; Pegg and Barnett 1988; Pegg and Barnett 1989), while another employs the anti-normal ordering of the SG operators (Luks and Perinova 1991; Luks and Perinova 1993).

The phase states $|\theta\rangle$ are defined as (Carruthers and Nieto 1968):

$$|\theta\rangle = \frac{1}{\sqrt{2\pi}} \sum_{n=0}^{\infty} e^{in\theta} |n\rangle \, . \tag{7.3}$$

These states are not orthogonal, due to the one-sided nature of the number spectrum (Carruthers and Nieto 1968), but they satisfy the resolution of the identity:

$$\int_0^{2\pi} d\theta |\theta\rangle\langle\theta| = \sum_{m,n=0}^{\infty} |m\rangle\langle n| \frac{1}{2\pi} \int_0^{2\pi} d\theta \exp\left[i(m-n)\theta\right] = \sum_{m,n=0}^{\infty} |m\rangle\langle n| \delta_{mn} = 1. \tag{7.4}$$

There are different operational methods for the measurement of phase distributions in one- and two-mode EM fields, many of which are based on phase states. However, since phase states are not orthogonal, it becomes important to understand the difference between '*projection-valued measure*' (PVM) and '*positive-operator valued measure*' (POVM) (Helstrom 1976).

7.1. Basic properties of PVM

A PVM measurement is obtained through a set of projection operators P_i which satisfy the following properties: 1) Hermiticity: $P_i = P_i^\dagger$; 2) Positive-semi-definiteness: $P_i \geq 0$; 3) Idempotency: $P_i^2 = P_i$; 4) Pairwise orthogonality; 5) Resolution of the identity: $\sum_{i=1}^{M} P_i = 1$. The probability of obtaining an outcome for the $i'th$ state from a state $|\psi\rangle$ is $\langle\psi|P_i|\psi\rangle$, and the post measurement is given by $|\psi^{(i)}{}_{post}\rangle = P_i|\psi\rangle / \sqrt{\langle\psi|P_i|\psi\rangle}$. For mixed states described by a density matrix $\hat{\rho}$ the probability of obtaining the $i'th$ outcome is given by $Tr(P_i\rho)$, and the post-measurement state becomes $\rho^{(i)}{}_{post} = P_i\rho P_i / Tr(P_i\rho)$.

7.2. Basic properties of POVM

In the case of a POVM, we omit the idempotent and orthogonality properties. However, POVM operators Q_i preserve the following features: 1) Hermiticity: $Q_i = Q_i^\dagger$; 2) Positive semi-definiteness: $Q_i \geq 0$; 3) Resolution of identity: $\sum_{i=1}^{M} Q_i = I$. The probability of obtaining the outcome is $Tr(Q_i\rho)$ but POVM does not specify the outcome due to the non-orthogonality. One can define a set of operators $Q_i = A_i^\dagger A_i$ where A_i is the Kraus operator, and this decomposition is the Kraus decomposition. To discriminate between non-orthogonal states, a PVM is generally insufficient, but POVM may succeed in certain cases. According to Neumark's theorem, any POVM defined on space U of a Hilbert space can be realized as a PVM performed on an extended Hilbert space.

The operators $|\theta\rangle\langle\theta|$ are not orthogonal and are therefore considered non-orthogonal POVM operators. In Barak and Ben-Aryeh (2005), we studied POVM phase distributions for one- and two-mode EM waves. The central idea is that such distributions can be related to the density matrix, obtained from quadrature probability distribution measurements. While in this work, Barak and Ben-Aryeh (2005) included extensive POVM calculations, in the following discussion, we focus only on the basic definitions of POVM phases.

For a single-mode EM, the quantum phase distribution is given as:

$$P(\theta) = Tr\{\hat{\rho}|\theta\rangle\langle\theta|\} = \sum_{m.n=0}^{\infty} \rho_{mn} \exp[i(m-n)\theta] \quad (7.5)$$

The wavefunction for a single-mode EM field can be given as $|\psi\rangle = \sum_{n=0}^{\infty} C_n |n\rangle$, where C_n is the amplitude corresponding to the photon number state $|n\rangle$ containing n photons. The corresponding quantum phase distribution for this state is given by:

$$P(\theta) = (2\pi)^{-1} \sum_{m,n=0}^{\infty} C_m^{*} C_n \exp[i(m-n)\theta] \quad (7.6)$$

For two EM modes, the quantum phase distribution may be described by using angular momentum basis (Barak and Ben-Aryeh 2005):

$$P_{POVM}(\theta) = (2\pi)^{-1} \sum_{j} \sum_{mm'} C^{*}_{jm} C_{jm'} \exp[i(m-m')\theta] \quad (7.7)$$

In Eq. (7.7), we employ the angular momentum basis $|j,m\rangle$ for each mode and restrict the summation to the cases where $j = j'$. This restriction ensures that the POVM phase depends only on $\exp[i(m-m')\theta]$, i.e., it arises from non-diagonal matrix elements in m, and not on non-diagonal elements in j. The present POVM angular momentum model is different from the one developed for the quantum phase by Luis and Sanchez-Soto (1993). Usually, the POVM description of the quantum phase is related to the use of phase states. However, it has been suggested to use a homodyne scheme to measure the phase distribution of quantum phase without using phase states (Vogel and Schleich 1991).

Pegg and Barnett (Barnett and Pegg 2007; Pegg and Barnett 1988; Pegg and Barnett 1989) suggested a solution to the quantum phase problem by redefining the phase states given by Eq. (7.3). They replaced the infinite sum with a finite sum which approaches infinity as:

$$|\theta\rangle = \lim \left\{ (s+1)^{-1/2} \sum_{n=0}^{s} \exp(in\theta) |n\rangle \right\}_{s \to \infty} . \quad (7.8)$$

Here, $|n\rangle$ are the ($s+1$) number states spanning an ($s+1$)-dimensional state space Ψ. The zero phase state is chosen as the state in which all the number states' amplitudes are equal. A limiting procedure is necessary to normalize the states. This procedure, proposed by Barnett and Pegg (2007) and developed in earlier works (Pegg and Barnett 1988; Pegg and Barnett 1989), operates entirely within states and operators in the (s+1)-dimensional space Ψ. Only after physical quantities such as expectation values are calculated, s tend to infinity. Phase states with values of θ differing by integer multiples of $2\pi/(s+1)$ are orthogonal. A complete set of ($s+1$) orthonormal phase states can be given as:

$$|\theta_m\rangle = \exp\left[i\hat{N}m2\pi/(s+1)\right]|\theta_0\rangle \ ; \ m = 0,1,2,\cdots,s \quad (7.9)$$

Here, we have used the unitary phase shift operator $\exp(i\hat{N}\gamma)$ which transforms θ to $\theta+\gamma$, and the initial phase θ_0 is chosen arbitrarily. The ($s+1$) values of θ_m are given by:

$$\theta_m = \theta_0 + \frac{2\pi m}{(s+1)} \quad (7.10)$$

The Hermitian phase operator is defined in terms of the phase state basis as:

$$\hat{\phi}_\theta = \sum_{m=0}^{s} \theta_m |\theta_m\rangle\langle\theta_m| = \theta_0 + \frac{2\pi}{s+1}\sum_{m-0}^{s} m|\theta_m\rangle\langle\theta_m|;$$
$$\hat{\phi}_\theta|\theta_m\rangle = \theta_m|\theta_m\rangle. \quad (7.11)$$

From the definition of the phase states given by Eq. (7.8), we express the projector $|\theta_m\rangle\langle\theta_m|$ in terms of the number state basis:

$$|\theta_m\rangle\langle\theta_m| = (s+1)^{-1}\sum_{n,n'}\exp\left[i(n'-n)\theta_m\right]|n'\rangle\langle n| \quad (7.12)$$

Pegg and Barnet (Barnett and Pegg 2007; Pegg and Barnett 1988; Pegg and Barnett 1989) constructed the unitary phase operator:

$$\exp\left(i\hat{\phi}_\theta\right) = |0\rangle\langle 1| + |1\rangle\langle 2| + \cdots + |s-1\rangle\langle s| + \exp\left[i(s+1)\theta_0\right]|s\rangle\langle 0| \quad (7.13)$$

Commuting phase operators $cos(\hat{\phi}_\theta)$ and $\sin(\phi_\theta)$ can be formed from the unitary phase operators $\exp(i\hat{\phi}_\theta)$, and its Hermitian conjugate $\exp(-i\hat{\phi}_\theta)$.

The exact relations between operators and states in the Ψ space and those in the conventional H space has been studied in great detail in the limit $s\to\infty$, by Vaccaro and Pegg (1993). They have shown that in this limit, the expectation values of Hermitian operators remain real, in agreement with a basic postulate of quantum mechanics. Moreover, they identified two types of limiting procedures in the transition from the Ψ space to H space. The strong limit holds for any power series in $\hat{a}$ and $a^{\dagger}$ as long as the algebra of bounded operators is preserved (Vaccaro and Pegg 1993):

$$\left(\lim \hat{B}_s\right)_{s\to\infty,strong}\left(\lim \hat{A}_s\right)_{s\to\infty,strong}|f\rangle=\lim\left(\hat{B}_s\hat{A}_s\right)_{s\to\infty,strong}|f\rangle, \qquad (7.14)$$

where $f\in H$. A sequence of operators $\hat{A}_s$ converges only weakly to $\hat{A}_W$ if:

$$\langle g|\hat{A}_s|f\rangle\to\langle g|\hat{A}_W|f\rangle \;\; as \;\; s\to\infty. \qquad (7.15)$$

Vaccaro and Pegg (1993) have shown that the Ψ space is consistent with H space in the weak limit.

Luks and Perinova (1991; 1993) defined the AN ordering of SG quantum phase operators as a procedure that places all raising operators $\exp(-i\hat{\phi})_{SG}$ to the right of all lowering operators $\exp(i\hat{\phi})_{SG}$. They claimed that this procedure should be applied to any products of SG phase operators. As a result, we obtain:

$$\left[\exp\left(-i\hat{\phi}\right)\exp\left(i\hat{\phi}\right)\right]_{AN}=\exp\left(i\hat{\phi}\right)\exp\left(-i\hat{\phi}\right)=1, \qquad (7.16)$$

where the subscript AN denotes anti-normal ordering. Justification for the anti-normal ordering of the SG phase operators is obtained from the Pegg-Barnet (PB) formalism (Vaccaro and Ben-Aryeh 1995), by which the SG operators are the weak limits on the infinite-dimensional Hilbert space H. Because the weak limit of a product of operators is not necessarily equal to the product of the two individual limits, it follows that the algebra of the PB operators will not be preserved under weak limits on H. However, we show that the AN ordering of the SG operators is precisely what is required to preserve the algebra for the weak limit on H.

In Brif and Ben-Aryeh (1994a; 1994b), we discussed several properties of SG phases obeying anti-normal ordering. Let us consider some of them in more detail. Specifically, we examine a class of operators that can be written in the form of a Fourier-like series in SG operators:

$$\hat{G}=\sum_{k=-\infty}^{\infty}G_k\left[\exp\left(-i\hat{\phi}\right)^k\right]_{AN} \quad ;$$

$$\left[\exp\left(-i\hat{\phi}\right)^k\right]_{AN}=\begin{cases}\left[\exp\left(-i\hat{\phi}\right)\right]^k & , \ k>0\\ \left[\exp\left(i\hat{\phi}\right)\right]^{|k|} & , \ k<0\\ 1 & , \ k=0\end{cases} . \tag{7.17}$$

Any quantum operator of the single-mode light field can be written in the number state representation by;

$$\hat{T}=\sum_{n,m=0}^{\infty}T_{nm}|n\rangle\langle m| \ . \tag{7.18}$$

Number representation matrix elements of an operator $\hat{G}$ given by Eq. (7.17) are:

$$G_{nm}\equiv\langle n|\hat{G}|m\rangle=\sum_{k=-\infty}^{\infty}G_k\langle n|\exp\left(-i\hat{\phi}\right)^k|m\rangle=\sum_{k=-\infty}^{\infty}G_k\delta_{k,n-m}=\hat{G}_{n-m} \ . \tag{7.19}$$

Then we get:

$$\hat{G}=\sum_{n.m=0}^{\infty}\hat{G}_{n-m}|n\rangle\langle m| \ . \tag{7.20}$$

We define the function $G(\theta)$ as the Fourier series with the coefficients $\hat{G}_k$ of Eq. (7.17):

$$G(\theta)=\sum_{k=-\infty}^{\infty}G_k\exp\left(-ik\theta\right) \ ; \ G_k=\frac{1}{2\pi}\int_{\theta_0}^{\theta_0+2\pi}d\theta G(\theta)e^{ik\theta} \ . \tag{7.21}$$

By substituting this expression for G_k into (7.20) we obtain:

$$\hat{G}=\frac{1}{2\pi}\sum_{n,m=0}^{\infty}\int_{\theta_0}^{\theta_0+2\pi}d\theta G(\theta)\exp\left[i(n-m)\theta\right]|n\rangle\langle m| \ . \tag{7.22}$$

By using Eq. (7.3) we get:

$$\hat{G} = \int_{\theta_9}^{\theta_0+2\pi} d\theta G(\theta)|\theta\rangle\langle\theta| \quad ; \quad \hat{G}^{\dagger} = \int_{\theta_9}^{\theta_0+2\pi} d\theta G^{*}(\theta)|\theta\rangle\langle\theta| \,. \tag{7.23}$$

Such a Fourier-like form for phase-related operators has also been discussed by Luks and Perinova (1991; 1993). Several important properties of AN ordering emerge from this formulation, e.g., as shown in Vaccaro and Ben-Aryeh (1995):

$$\left[\hat{G}\hat{F}\right]_{AN} = \int_{\theta_0}^{\theta_0+2\pi} d\theta G(\theta)F(\theta)|\theta\rangle\langle\theta| \,. \tag{7.24}$$

References

Barak, R., and Y. Ben-Aryeh. 2005. "Non-Orthogonal Positive Operator Valued Measure Phase Distributions and One- and Two-Mode Electromagnetic Fields." *Journal of Optics B: Quantum and Semiclassical Optics* 7: 123–130. https://doi.org/10.1088/1464-4266/7/5/001

Barnett, S. M., and D. T. Pegg. 2007. "On the Hermitian Optical Phase Operator." *Journal of Modern Optics* 36: 7–19. https://doi.org/10.1080/09500348914550021

Brif, C., and Y. Ben-Aryeh. 1994. "Antinormally Ordering of Susskind-Glogower Quantum Phase Operators." *Physical Review A* 50: 2727–2731. https://doi.org/10.1103/PhysRevA.50.2727

Brif, C., and Y. Ben-Aryeh. 1994. "Phase State Representation in Quantum Optics." *Physical Review A* 50: 3505–3515. https://doi.org/10.1103/PhysRevA.50.3505

Carruthers, P., and M. M. Nieto. 1968. "Phase and Angle Variables in Quantum Mechanics." *Reviews of Modern Physics* 40: 411–440. https://doi.org/10.1103/RevModPhys.40.411

Gerry, C., and P. Knight. 2004. *Introductory Quantum Optics*. Cambridge: Cambridge University Press. https://doi.org/10.1017/CBO9780511791239

Helstrom, C. W. 1976. *Quantum Detection and Estimation Theory*. New York: Academic Press.

Luis, A., and L. L. Sanchez-Soto. 1993. "Phase Difference Operator." *Physical Review A* 48: 4702–4711. https://doi.org/10.1103/PhysRevA.48.4702

Luks, A., and V. Perinova. 1991. "Extended Number State Basis and Number Phase Intelligent States. I. Mapping and Operator Ordering Approach." *Czechoslovak Journal of Physics* 41: 1205–1220. https://doi.org/10.1007/BF01613540

Luks, A., and V. Perinova. 1993. "Ordering of 'Ladder' Operators, the Wigner Function for Number Phase, and the Enlarged Hilbert Space." *Physica Scripta* T48: 94–98. https://doi.org/10.1088/0031-8949/1993/T48/014

Pegg, D. T., and S. M. Barnett. 1988. "Unitary Phase Operator in Quantum Mechanics." *Europhysics Letters* 6: 483–487. https://doi.org/10.1209/0295-5075/6/6/002

Pegg, D. T., and S. M. Barnett. 1989. "Phase Operator of the Quantized Single Mode Electromagnetic Field." *Physical Review A* 39: 1665–1675. https://doi.org/10.1103/PhysRevA.39.1665

Schleich, W. 2015. *Quantum Optics in Phase Space*. Weinheim: Wiley.

Susskind, L., and J. Glogower. 1964. "Quantum Mechanical Phase and Time Operator." *Physics* 1: 49–61. https://doi.org/10.1103/PhysicsPhysiqueFizika.1.49

Vaccaro, J. A., and D. T. Pegg. 1993. "Consistency of Quantum Description of Phase." *Physica Scripta* T48: 22–27. https://doi.org/10.1088/0031-8949/1993/T48/003

Vaccaro, J. A., and Y. Ben-Aryeh. 1995. "Antinormally Ordering of Phase Operators and the Algebra of the Weak Limit." *Optics Communications* 113: 427–431. https://doi.org/10.1016/0030-4018(94)00624-4

Vogel, W., and W. Schleich. 1991. "Phase Distribution of a Quantum Phase Without Using Phase States." *Physical Review A* 44: 7642–7647. https://doi.org/10.1103/PhysRevA.44.7642

CHAPTER 8

BERRY AND PANCHARATNAM GEOMETRIC PHASES, METRIC, AND GEODESIC

In the following discussion, we explain some basic concepts from geometric phases theory, specifically, *parallelisms, connections, and curvatures* (Shapere and Wilczek 1989; Bohm et al. 2003; Chruściński and Jamiołkowski 2004; Ben-Aryeh 2004).

The phase of a wave function representing a particle with charge e at a spatial point $\vec{x}$ is said to be *parallel* to the phase at another point $\vec{x}+d\vec{x}$, if the local values of the phase at these two points differ by an amount $eA_\mu dx^\mu$. Here, $\vec{A}(\vec{x})$ is the EM vector potential at point $\vec{x}$, and $A_\mu(\vec{x})$ denoted its component along the coordinate x_μ. Accordingly, the wave function at point $\vec{x}+d\vec{x}$ is transformed to:

$$\Psi(\vec{x}+d\vec{x}) = \Psi(\vec{x}) + eA_\mu dx^\mu \quad ; \quad \mu = 1,2,3 \, . \tag{8.1}$$

Eq. (8.1) is referred to as the *connection*. It is based on the previously discussed concept of *parallelism*, which incorporates the use of *vector potential.* However, the vector potential itself is defined only up to a *gauge transformation, expressed by an additional term* $e\alpha(\vec{x})$ given as:

$$\Psi'(\vec{x}) \rightarrow \exp\left[ie\alpha(\vec{x})\right]\Psi(\vec{x}) \, . \tag{8.2}$$

Then, the phase at a neighboring point changes as:

$$\Psi'(\vec{x}+d\vec{x}) = \exp\left[ie\left\{\alpha(\vec{x}) + \partial_\mu\alpha(\vec{x})\,dx^\mu\right\}\right]\Psi(\vec{x}+d\vec{x}) \, . \tag{8.3}$$

To maintain *parallelism* between two neighboring points in the transformed system, the vector potential A'_μ is given as $[eA_\mu + \partial_\mu\alpha(\vec{x})]$.

Although $A_\mu(\vec{x})$ is, to some extent, arbitrary, due to the gauge relation of Eq. (8.3), this arbitrariness vanishes for a closed circuit as:

$$e\oint \partial_\mu \alpha(\vec{x}) dx^\mu = 0 \quad (8.4)$$

For a parallel transport, the phase of the wavefunction will be altered over a closed circuit according to:

$$\Delta\psi(C) = e\oint A_\mu(\vec{x}) dx^\mu = e\iint_S f_{\mu\nu} d\sigma^{\mu\nu} \; ;$$

$$f_{\mu\nu} = \partial_\nu A_\mu(\vec{x}) - \partial_\mu A_\nu(\vec{x}). \quad (8.5)$$

This equation is derived by the Stokes theorem. Here, $f_{\mu\nu}$ is the EM field tensor, which in topological terms is referred to as the *curvature*. The surface integral $e\iint_s f_{\mu\nu} d\sigma^{\mu\nu}$ is taken over the area bounded by the closed circuit.

The above U(1) transformations can be generalized to SU(2) transformations (Wilczek and Zee 1984; Zee 1988), expressed as:

$$\vec{\psi}'(\vec{x}) = S(\vec{x})\vec{\psi}(\vec{x}) \quad (8.6)$$

Here S is a 2×2 unitary matrix with unit determinant. The state $\vec{\psi}(\vec{x}+d\vec{x})$ is said to be *parallel* to $\vec{\psi}(\vec{x})$ under the condition:

$$\vec{\Psi}(\vec{x}+d\vec{x}) = \exp\left[igA_\mu(\vec{x})dx^\mu\right]\vec{\Psi}(\vec{x}), \quad (8.7)$$

where the generalized potentials are given by 2×2 matrices, which *depend on the special physical system* (Wilczek and Zee 1984; Zee 1988). Since, in the general case, the matrices do not commute, the *path-ordered phase change for a closed loop C is given in the non-Abelian case* by:

$$\vec{\psi}_C(\vec{x}) = P\left[ig\oint_C A_\mu(\vec{x}) dx^\mu\right]\vec{\psi}_{in}(\vec{x}) \quad (8.8)$$

In this non-Abelian case, path ordering must be used when manipulating Eq. (8.8) (Hong-Mo and Tsun 1993), denoted by the symbol P, and the resulting equations might become quite complicated.

Aharonov and Bohm (AB) (Aharonov and Bohm 1959) proposed that a charged particle passing around a long solenoid experiences a quantum phase shift as a result of the enclosed magnetic flux, even though the electrons move through a region free of electric and magnetic fields. This AB effect was later verified experimentally (Holstein 1991; Peshkin and Lipkin 1995). These treatments, however, overlook the fact that the AB effect is fundamentally a three-system effect involving the electron, the solenoid, and the EM fields. Although there is no force acting on the electron, there is a force acting on the mechanical solenoid. This result followed from our analysis (Ben-Aryeh and Mann 2000), which considered the inequality of action and reaction forces between two bodies within a three-system description of the AB effect. In our analysis, we have shown that interference occurs between the electric field produced by the electron inside the solenoid and its magnetic field. The key point in our analysis is that the force acting on the solenoid is equal in magnitude and opposite in sign to the rate of change of momentum of the EM field. Our calculations (Ben-Aryeh and Mann 2000) showed that the integral of the energy interference $[\vec{E}_{el}(\vec{r},\vec{r}\,')\times\vec{B}_{Sol}]d^3r'$ over the closed trajectory $\vec{r}$ of the electron is proportional to the AB phase shift. Here, $\vec{E}_{el}(\vec{r},\vec{r}\,')$ is the electric field at point $\vec{r}\,'$ in the solenoid, produced by the electron located at point $\vec{r}$, and $\vec{B}_{Sol}$ is the constant magnetic field in the solenoid.

Due to the importance of Berry's article on geometric phases (Berry 1984), which marked a breakthrough in this field, we summarize here some of its essential results. Using the adiabatic approximation, Berry considered physical systems transported around a closed path in parameter space $\vec{R}(t)$ with a Hamiltonian $H(\vec{R}(t))$. For simplicity, $\vec{R}$ was assumed to be three-dimensional. Suppose a quantum system is initially in an eigenstate of the Hamiltonian and evolves with time under slowly varying parameters. In that case, the adiabatic approximation guarantees that it remains in an eigenstate of the instantaneous Hamiltonian throughout the process. Assuming an initial eigenstate of the Hamiltonian:

$$H(\vec{R})\left|n,\vec{R}\right\rangle = E_n\left(\vec{R}\right)\left|n,\vec{R}\right\rangle, \tag{8.9}$$

where n denotes nondegenerate eigenstate of the Hamiltonian, and $\vec{R}$ represents parameters that vary slowly along a closed curve C in parameter space in time T, the geometric phase acquired by the system is given by:

$$\gamma_n(C) = i\oint d\vec{R}\cdot\langle n,\vec{R}|\vec{\nabla}_{\vec{R}}|n,\vec{R}\rangle . \tag{8.10}$$

This phase is given in addition to the dynamical phase, which is given by:

$$\phi_{dyn} = -\int_0^T E_n(\tau)\,d\tau . \tag{8.11}$$

After further calculations and applying Stokes' theorem, a more special form for the geometric phase was obtained (Berry 1984):

$$\gamma_n(C) = -\iint_C d\hat{S}\cdot\vec{V}_n(\vec{R}) , \tag{8.12}$$

where $d\vec{S}$ denotes the area element in $\vec{R}$ parameter space, and $\vec{V}_n(\vec{R})$ is given as (Berry 1984):

$$V_n(C) = \frac{\left\{\operatorname{Im}\sum_{n\neq m}\left[\langle n(\vec{R})|\nabla_{\vec{R}}H(\vec{R})|m(\vec{R})\rangle\langle m(\vec{R})|\nabla_{\vec{R}}H(\vec{R})|n(\vec{R})\rangle\right]\right\}}{\left[E_m(\vec{R})-E_n(\vec{R})\right]^2} . \tag{8.13}$$

Berry (1984) analyzed the case in which the circuit C lies close to a point with a degeneracy. He considered the most common situation, where the degeneracy involves only two states, denoted by $(+)$ and (–), and where $E_+(\vec{R})\geq E_-(\vec{R})$. For a circuit that begins at the $(+)$ state, he obtained, after some calculations, the following result:

$$V_n(\vec{R}) = V_+(\vec{R}) = \frac{\vec{R}}{2R^3} , \tag{8.14}$$

where $\gamma_+(C)$ is the geometric phase acquired. This phase corresponds to the flux through the circuit *C* of the magnetic field of a monopole with strength -1/2, located at the degeneracy $\vec{R}=0$, where $V_-(\vec{R})=-V_+(\vec{R})$; $\gamma_-(C)=-\gamma_+(C)$, and

$$\exp\left(i\gamma_\pm\left(C\right)\right)=\exp\left[\pm\frac{1}{2}\Omega\left(C\right)\right], \tag{8.15}$$

where $\Omega\left(C\right)$ is the solid angle subtended by the circuit C at the point of degeneracy (Berry 1984).

Berry (1984) also treated the case of a particle with spin S (either integer or half-integer) interacting with a magnetic field $\vec{B}$ governed by the Hamiltonian:

$$H\left(\vec{B}\right)=\kappa\hbar\vec{B}\cdot\vec{s} \quad ; \quad E_n=\kappa\hbar\vec{B}n, \tag{8.16}$$

where κ is a constant involving the gyromagnetic ratio, and $\vec{s}$ is the spin vector operator with $2s+1$ eigenvalues n ranging from $-s$ and $+s$. The phase change $\gamma_n(C)$ was calculated for the case in which $\vec{B}$ is slowly varying (and hence the spin is rotated) around a circuit C, obtaining the result:

$$\vec{V}_n\left(B\right)=\frac{n\vec{B}}{B^3}. \tag{8.17}$$

The components of $\vec{B}$ correspond to the parameters $\vec{R}$ used in the previous analysis. The geometric phase is then given by:

$$\exp\left[i\gamma_n\left(C\right)\right]=\exp\left[-in\Omega\left(C\right)\right], \tag{8.18}$$

where $\Omega(C)$ is the solid angle that C subtends at $\vec{B}=0$.

Aharonov and Anandan (1987) removed the dynamical phase factor by defining a new state:

$$\Phi(t)=\exp\left[i\int_0^t h(t')dt'\right]|\Psi(t)\rangle \quad , \tag{8.19}$$

where

$$h(t')=\langle\psi|\psi\rangle^{-1}\,\mathrm{Re}\langle\psi(t')|H(t')|\Psi(t')\rangle \, . \tag{8.20}$$

The geometric phase, now isolated from any dynamical contribution, is given by:

$$\gamma_{geom}=i\oint dt\left\langle\Phi\left(\vec{R},t\right)\right|\frac{d}{dt}\left|\Phi\left(\vec{R},t\right)\right\rangle \, . \tag{8.21}$$

8.1. Pancharatnam phase and noncyclic evolution

Consider two normalized nonorthogonal Hilbert states $|A\rangle$ and $|B\rangle$, and assume further that $|A\rangle$ is subjected to a $U(1)$ phase shift $e^{i\varphi}$ (Pancharatnam 1956). The resulting interference pattern is determined by:

$$I=\left|e^{i\varphi}|A\rangle+|B\rangle\right|^2=2+2|\langle A|B\rangle|\left[\cos\phi-arg\langle A|B\rangle\right] \, . \tag{8.22}$$

The maximum is obtained at the Pancharatnam phase $\phi_0=\arg\langle A|B\rangle$. This phase is reduced to the $U(1)$ case whenever $|B\rangle=e^{i\theta}|A\rangle$, as it then yields $arg\langle A|B\rangle=\theta$. The phase difference between the state $|A\rangle$ and the state $|B\rangle$ is well-defined and can be observed using interferometric methods. In a noncyclic evolution, if the final wave $|B\rangle$ is superimposed on the initial state $|A\rangle$, only the component $\langle A|B\rangle|A\rangle$ along $|A\rangle$ interferes with it. The "*Pancharatnam connection*" defines the phase between $|B\rangle$ and $|A\rangle$ as $\phi_0=\arg\langle A|B\rangle$. The interference amplitude $|\langle A|B\rangle|$ differs from unity if the evolution is non-cyclic, and this difference leads to a reduction in the visibility of the interference pattern. The Pancharatnam phase becomes indeterminate if $|B\rangle$ is orthogonal to $|A\rangle$.

The use of Pancharatnam phases in non-cyclic evolution raises the problem of gauge invariance. Under the adiabatic approximation, the non-cyclic geometric phase is defined as:

$$\gamma_n(C) = \arg\left\langle n, \vec{R}(0) \middle| n, \vec{R}(t) \right\rangle + i \int_{\vec{R}(0)}^{\vec{R}(t)} d\vec{R} \cdot \left\langle n, \vec{R} \middle| \vec{\nabla}_{\vec{R}} \middle| n, \vec{R} \right\rangle, \tag{8.23}$$

where the initial state $\left|n, \vec{R}(0)\right\rangle$ at time $t = 0$ evolves into the state $\left|n, \vec{R}(t)\right\rangle$ at time t. For $\left|n, \vec{R}(t)\right\rangle \neq \left|n, \vec{R}(0)\right\rangle$ this development is non-cyclic. We then have the following relation:

$$\begin{gathered} \operatorname{Re}\left\langle n, \vec{R} \middle| \vec{\nabla}_{\vec{R}} \middle| n, \vec{R} \right\rangle = 0; \\ \left\langle n, \vec{R} \middle| \vec{\nabla}_{\vec{R}} \middle| n, \vec{R} \right\rangle = i \operatorname{Im}\left\langle n, \vec{R} \middle| \vec{\nabla}_{\vec{R}} \middle| n, \vec{R} \right\rangle. \end{gathered} \tag{8.24}$$

Therefore, we can express the geometric phase as:

$$\gamma_n(C) = \arg\left\langle n, \vec{R}(0) \middle| n, \vec{R}(t) \right\rangle - \operatorname{Im} \int_{\vec{R}(0)}^{\vec{R}(t)} d\vec{R} \cdot \left\langle n, \vec{R} \middle| \vec{\nabla}_{\vec{R}} \middle| n, \vec{R} \right\rangle. \tag{8.25}$$

The crucial point here is that $\gamma_n(C)$ is gauge invariant. In Eq. (8.25), any gauge transformation introduces phase shifts that are equal in magnitude and opposite in sign in the two terms on the right-hand side of this equation.

8.2. Metric and geodesic for quantum geometric phase

A *fiber bundle* is defined by the following set of requirements:

a) We have a certain manifold M with a *base space* X.
b) We have another manifold F, called *fiber*, which is defined over the base space X.

c) The total space E includes both the base space X and the *fiber* F, along with a *projection* π of E onto $X : \pi E \to X$. A fiber bundle E over M with a fiber F is covered with a set of local neighborhoods $\{U_i\}$ where in each U_i, the bundle neighborhood of E is described by the product $U_i \times F$.

d) The bundle E is specified by a set of *transition functions* $\{\phi_{ij}\}$ which transform the manifolds between two neighborhoods, $U_i \cap U_j$.

e) There is a set of open neighborhoods U_α covering X, where for each U_α:

$$\pi^{-1}U_\alpha = U_\alpha \times F \quad .(8.26)$$

Although the *local topology* of the bundle is trivial, the *global topology* of the *fiber bundle* might be quite complicated as a consequence of nontrivial transition functions.

f) If $x \in X$ i.e., if x is a certain element in the base space X, then $\pi^{-1}(x)$ is the *fiber f* over x, i.e., f is the fiber corresponding to the point x. An element of the bundle E can be written as $(x, f), f \in F, x \in U_\alpha$, where U_α is a certain neighborhood of X.

g) If $(x, f) \in U_\alpha \times F$ and $(x', f') \in U_\beta \times F$, then $(x, f) \sim (x', f')$ if $x = x'; \phi_{\alpha\beta}(x) f = f'$ where $\sim$ denotes an equivalence relation.

We find that the following 'connection' for the Hilbert space metric is closely related to fiber bundle theory. The conventional approach for 'parallelism' in the Schrödinger equation is to use the 'connection' introduced by Garcia de Polavieja and Sjöqvist (1998):

$$\langle \Phi | d\Phi \rangle = 0 \quad ; \quad \langle \Phi | d\Phi / dt \rangle = 0 \quad (8.27)$$

Here, we eliminate the dynamical phases for $|\Phi\rangle$, by applying the transformations given in Eqs. (8.19-8.21).

Since $\langle\Phi|\Phi\rangle = 1 \rightarrow \langle d\Phi/dt|\Phi\rangle + \langle\Phi|d\Phi/dt\rangle = 0$, it follows that $\langle\Phi|d\Phi/dt\rangle$ is necessarily imaginary due to normalization. The equality in Eq. (8.27) follows from the condition of parallelism in the Schrödinger equation, where this equation represents the definition of *connection*. Physically, this means that $|\Phi(t+\Delta t)\rangle$ is in phase with $|\Phi(t)\rangle$, so that any change in phase should be attributed to the 'fiber' and not to the base functions. The underlying idea is that phases do not follow the transitive law. That is, if system A is in phase with system B, and system B is in phase with system C, it does not follow that system A is in phase with system *C*, in the general case. The idea of geometric phases is that although locally there is no change in the geometric phase, there can be a global phase change that depends on the geometry of the physical system. The connection used in Eq. (8.27) is analogous to the one introduced earlier by Pancharatnam (1956), who defined two light beams to be in phase if the intensity of their sum is maximal.

The distance between two quantum states $|\psi(s)\rangle$ and $|\psi(s')\rangle$, representing a quantum system evolving with respect to parameter s, can be given as:

$$dL^2 = 1 - \left|\langle\psi(s)|\psi(s')\rangle\right|^2 . \tag{8.28}$$

The use of Eq. (8.28) is justified since if the ray $|\psi(s)\rangle$ is orthogonal to the ray $|\psi(s')\rangle$ then the distance is equal 1, while if $|\psi(s')\rangle$ is equal to $|\psi(s)\rangle$, the distance vanishes.

Assuming two close wavefunctions $|\psi(s)\rangle$ and $|\psi(s+ds)\rangle$, we obtain:

$$\langle\psi(s)|\psi(s+ds)\rangle = 1 + ds\langle\psi|\frac{d}{ds}|\psi\rangle + \frac{1}{2}ds^2\langle\psi|\frac{d^2}{ds^2}|\psi\rangle . \tag{8.29}$$

We neglect terms of order ds^3 or higher. Given the continuous 'curve' C of the wavefunction, $\Psi(s)$, which varies continuously as a function of the parameter s, from s_1 to s_2, one obtains, after some calculations, the functional (Mukunda and Simon 1993):

$$L(C)=\int_{s(1)}^{s(2)}\left[\left\langle\frac{d\Psi(s)}{ds}\middle|\frac{d\Psi(s)}{ds}\right\rangle-\left\langle\frac{d\Psi(s)}{ds}\middle|\Psi(s)\right\rangle\left\langle\Psi(s)\middle|\frac{d\Psi(s)}{ds}\right\rangle\right]^{1/2}ds \quad . \tag{8.30}$$

The geodesic curve is obtained by extremizing this functional in Eq. (8.30). Mukunda and Simon (1993) showed that any 'curve' composed of a normalized wavefunction, which varies continuously with respect to the parameter s, and which satisfies the following equations:

$$\left|\frac{d^2\Psi(s)}{ds^2}\right\rangle=-\left\langle\frac{d\Psi(s)}{ds}\middle|\frac{d\Psi(s)}{ds}\right\rangle\left|\Psi(s)\right\rangle, \tag{8.31}$$

$$\left\langle\Psi(s)\middle|\frac{d\Psi(s)}{ds}\right\rangle=0, \tag{8.32}$$

is a geodesic curve. The path

$$\left|\psi(s)\right\rangle=\left|A\right\rangle\cos(s)+\frac{\left|B\right\rangle-\left|A\right\rangle\left\langle A\middle|B\right\rangle}{\left(1-\left|\left\langle B\middle|A\right\rangle\right|^2\right)^{1/2}}\sin(s) \quad , \tag{8.33}$$

where $|A\rangle$ and $|B\rangle$ are the initial and the final wavefunctions, respectively, is a geodesic curve, as it satisfies Eqs. (8.31-8.32).

I consider the above geodesic curve analysis as quantum generalization of the classical definition of geodesic curve by parallel transport (Ben-Aryeh 2004). There is an extensive body of work on geometric phases in various physical systems, some of which was reviewed in our earlier article (Ben-Aryeh 2004). In the following discussion, we analyze special geometric phases related to the Cartan subalgebra of atomic systems, a topic explored in detail in our previous works (Ben-Aryeh 2002; Ben-Aryeh 2003; Ben-Aryeh 2007).

8.3. Geometric phases of two-level and three-level states

We use group theory to analyze geometric phases of nondegenerate two-level and three-level systems, based on a special representation of the time development operator (Ben-Aryeh 2002):

$$U(t)=\xi(t)h(t), \tag{8.34}$$

where $h(t)$ includes the operators belonging to the Cartan subalgebra, and $\xi(t)$ includes all other operators of the group.

For the two-level system, $U(t)$ operates on the two-dimensional vector:

$$|\psi(t)\rangle=\begin{pmatrix} C_1(t) \\ C_2(t) \end{pmatrix}, \tag{8.35}$$

where $C_1(t)$ and $C_2(t)$ are the amplitudes for the two levels. For this case, the Cartan subalgebra includes only the Pauli spin matrix σ_z, so that $h(t)$ is given as:

$$h(t)=\exp\left(-i\theta(t)\sigma_z\right), \tag{8.36}$$

where $\theta(t)$ is the Cartan geometric phase. The operator $\xi(t)$ includes other operators of the SU(2) group, i.e., the Pauli spin operators σ_x and σ_y.

For the non-degenerate atomic system, the unitary operator $U(t)$ operates on the three-dimensional vector:

$$|\psi(t)\rangle=\begin{pmatrix} C_1(t) \\ C_2(t) \\ C_3(t) \end{pmatrix}, \tag{8.37}$$

where $C_1(t)$, $C_2(t)$, and $C_3(t)$ are the amplitudes corresponding to the three atomic levels. The Cartan subalgebra operators, in the three-dimensional matrix representation for the three-level system, are given as (Ben-Aryeh 2002; Ben-Aryeh 2003; Ben-Aryeh 2007; Gilmore 1974):

$$H_1 = \begin{pmatrix} 1 & 0 & 0 \\ 0 & -1 & 0 \\ 0 & 0 & 0 \end{pmatrix}; \; H_2 = \begin{pmatrix} 1 & 0 & 0 \\ 0 & 1 & 0 \\ 0 & 0 & -2 \end{pmatrix}. \tag{8.38}$$

By using the commuting operators H_1 and H_2, the time development of $h(t)$, of the SU(3) group, is given as (Ben-Aryeh 2002; Ben-Aryeh 2003; Ben-Aryeh 2007; Gilmore 1974):

$$h(t) = \exp\left[-i\theta_1(t)H_1 t - i\theta_2(t)H_2 t\right], \tag{8.39}$$

where $\theta_1(t)$ and $\theta_2(t)$ are the Cartan geometric phases, and $\xi(t)$ includes all other SU(3) operators.

According to the Schrödinger equation,

$$H(t)\left|n(t)\right\rangle = i\dot{U}(t)\left|n(0)\right\rangle, \tag{8.40}$$

where the dot denotes the time derivative. Here, $\left|n(0)\right\rangle$ is any of the initial states ($\left|n(0)\right\rangle = \left|1\right\rangle$, $\left|2\right\rangle$, or $\left|3\right\rangle$). Substituting Eq. (8.34) in Eq. (8.40), we get:

$$-iH\xi(t)h(t) = \frac{d\xi(t)}{dt}h(t) + \xi(t)\frac{dh(t)}{dt}, \tag{8.41}$$

where the term $\left|n(0)\right\rangle$ is cancelled from both sides of the equation. Multiplying both terms on the right-hand side by $h^{-1}(t)$, and on the left-hand side by $\xi^{-1}(t)$, we get:

$$-i\xi^{-1}(t)H(t)\xi(t) = \xi^{-1}(t)\frac{d\xi}{dt} + \frac{dh(t)}{dt}h^{-1}(t). \tag{8.42}$$

According to Eq. (8.39), we then find for the three-level system:

$$\frac{dh(t)}{dt}h^{-1}(t) = -i\frac{d\theta_1(t)}{dt}H_1 - i\frac{d\theta_2(t)}{dt}H_2, \tag{8.43}$$

while for the two-level system, we get:

$$\frac{dh(t)}{dt}h^{-1}(t) = -i\frac{d\theta(t)}{dt}H \;\; ; \;\; H = \sigma_z . \tag{8.44}$$

On the left-hand side of Eq. (8.42), a 3×3 matrix $\xi(t)$ will diagonalize H for the three-level system, and a 2×2 matrix $\xi(t)$ will diagonalize H for the two-level system, with real eigenvalues. These eigenvalues are given here under the adiabatic approximation. On the right-hand side of Eq. (8.42), we obtain the imaginary terms as:

$$\xi^{-1}(t)\frac{d\xi}{dt} = -\frac{dh(t)}{dt}h^{-1}(t) \quad , \tag{8.45}$$

where the right-hand side of Eq. (8.45) follows from the imaginary components of Eq. (8.43) or Eq. (8.44) for the three-level or for the two-level state, respectively,

8.4. Explicit calculation of the geometric phase for a two-level system

For a two-level system, the Cartan subalgebra consists of only one generator, i.e., the Pauli operator σ_z. For this system, we have the following:

a) The Hamiltonian $H = H_0 + V(t)$ in the rotating frame is given by:

$$H = H_0 + V(t) = \begin{pmatrix} \Delta E & \kappa(t) \\ \kappa^{*}(t) & -\Delta E \end{pmatrix} \quad , \tag{8.46}$$

where $2\Delta E$ is the deviation from resonance.

b) The time development operator $h(t)$ is given by:

$$h(t) = \exp\left[-i\theta_1(t)\sigma_z\right] \quad , \tag{8.47}$$

where $\theta(t)$ is the time-dependent phase parameter.

c) The time derivative of the phase $\Delta\theta$ is given by:

$$2\frac{d\Delta\theta}{dt}=-iTr\left[\sigma_z\xi^{-1}(t)\frac{d\xi}{dt}\right]. \quad (8.48)$$

Here, $2d\Delta\theta/dt$ represents the phase shift, in addition to the dynamical phase given by $Tr[\sigma_z\xi^{-1}(t)H(t)\xi(t)]$.

Under the adiabatic approximation, the dynamical phase factor is given by $E_1(t)-E_2(t)$, where $E_1(t)$ and $E_2(t)$ are the eigenvalues of the Hamiltonian given in Eq. (8.46).

The time development operator $\xi(t)$ is then written as:

$$\xi(t)=\exp\left[i\alpha(t)\sigma_x+i\beta\sigma_y\right]=$$

$$\exp\begin{pmatrix}0 & \phi(t)e^{i\delta(t)}\\ -\phi(t)e^{-i\delta(t)} & 0\end{pmatrix}=\begin{pmatrix}\cos\phi(t) & \sin\phi(t)e^{i\delta(t)}\\ -\sin\phi(t)e^{-i\delta(t)} & \cos\phi(t)\end{pmatrix}, \quad (8.49)$$

where we have defined:

$$\beta(t)+i\alpha(t)=\phi(t)e^{i\delta(t)}, \quad (8.50)$$

and σ_x, and σ_y are Pauli spin operators.

The matrix $\xi^{-1}(t)$ is given by:

$$\begin{pmatrix}\cos\phi(t) & -\sin\phi(t)e^{i\delta(t)}\\ \sin\phi(t)e^{-i\delta(t)} & \cos\phi(t)\end{pmatrix}. \quad (8.51)$$

Integrating the geometric phase from the initial time $t=0$ to time T, we get:

$$\theta_{Geom}=-i\int_0^T Tr\left[\sigma_z\xi^{-1}\frac{d\xi(t)}{dt}\right]. \quad (8.52)$$

The parameters $\phi(t)$ and $\delta(t)$ are fixed by the requirement that $\xi^{-1}(t)H(t)\xi(t)$ is diagonal, where $\hat{H}$ was given by Eq. (8.46). The non-diagonal terms in Eq. (8.52) vanish under the conditions:

$$\tan(2\phi(t)) = -\frac{|\kappa(t)|}{\Delta E(t)}; \quad \delta(t) = \arg(\kappa(t));$$

$$\cos(2\phi(t)) = \frac{\Delta E(t)}{\sqrt{\Delta E(t)^2 + |\kappa|^2}}; \ \sin(2\phi(t)) = \frac{-|\kappa(t)|}{\sqrt{\Delta E(t)^2 + |\kappa|^2}}. \tag{8.53}$$

Using these conditions, we get:

$$\xi^{-1}(t)H(t)\xi(t) = \begin{pmatrix} \sqrt{(\Delta E)^2 + |\kappa|^2} & 0 \\ 0 & -\sqrt{(\Delta E)^2 + |\kappa|^2} \end{pmatrix} = \begin{pmatrix} E_1 & 0 \\ 0 & E_2 \end{pmatrix}, \tag{8.54}$$

where $\pm\sqrt{(\Delta E)^2 + |\kappa|^2}$ are the two energy eigenvalues under the adiabatic approximation. By using Eqs. (8.49) and (8.52), we get the expression for the geometric phase under the adiabatic approximation:

$$\theta_{Geom} = -\int_0^T \sin^2\phi d\delta \ . \tag{8.55}$$

The ordinary Berry phase for a closed circuit, associated with the upper level of a two-level system, is equal in magnitude but opposite in sign relative to that of the lower level. Therefore, the Cartan geometric phase of the two-level system over a closed circuit is twice the value of the ordinary Berry phase. While Eq. (8.55) can also be applied to an open circuit, one needs to guarantee gauge invariance, particularly in the context of the Pancharatnam phase. As an alternative to the approach presented here, the geometric phase can also be calculated using complex variables, including the application of the complex Stokes theorem (Ben-Aryeh 2003; Ben-Aryeh 2007).

Explicit calculations of the geometric phases for the three-level atomic system were carried out using the method described above (Ben-Aryeh 2007). These calculations turn out to be quite complex, as in the general case, the matrix $\xi(t)$ depends on six generators, and the resulting solutions involve solving cubic equations.

References

Aharonov, Y., and J. Anandan. 1987. "Phase Change during a Cyclic Quantum Evolution." *Physical Review Letters* 58: 1593. https://doi.org/10.1103/PhysRevLett.58.1593

Aharonov, Y., and D. Bohm. 1959. "Significance of Electromagnetic Potentials in the Quantum Theory." *Physical Review* 115: 485. https://doi.org/10.1103/PhysRev.115.485

Ben-Aryeh, Y. 2002. "Berry Phase and Interference Effects in Three-Level Atoms." *Journal of Modern Optics* 49: 207. https://doi.org/10.1080/09500340110087237

Ben-Aryeh, Y. 2003. "Berry Phase Effects in Two-Level and Three-Level Atoms." *Optics and Spectroscopy* 94: 724–729. https://doi.org/10.1134/1.1576842

Ben-Aryeh, Y. 2004. "Berry and Pancharatnam Topological Phases of Atoms and Optical Systems." *Journal of Optics B: Quantum and Semiclassical Optics* 6: R1–R18. https://doi.org/10.1088/1464-4266/6/4/R01

Ben-Aryeh, Y. 2007. "Geometric Phases of Two- and Three-Level Atomic Systems Related to Group Theory." *Journal of Modern Optics* 54: 565. https://doi.org/10.1080/09500340601004951

Ben-Aryeh, Y., and A. Mann. 2000. "Aharonov-Bohm Solenoid Effect Related to the Energy of Interference." *Journal of the Optical Society of America B* 17: 2052. https://doi.org/10.1364/JOSAB.17.002052

Berry, M. V. 1984. "Quantal Phase Factors Accompanying Adiabatic Changes." *Proceedings of the Royal Society A* 392: 45. https://doi.org/10.1098/rspa.1984.0023

Bohm, A., A. Mostafazadeh, H. Koizumi, Q. Niu, and J. Zwanziger. 2003. *The Geometric Phase in Quantum Systems*. Berlin: Springer. https://doi.org/10.1007/978-3-662-10333-3

Chruściński, D., and A. Jamiołkowski. 2004. *Geometric Phases in Classical and Quantum Mechanics*. Boston: Birkhäuser. https://doi.org/10.1007/978-0-8176-8176-0

Garcia de Polavieja, G., and E. Sjöqvist. 1998. "Extending the Quantal Adiabatic Theorem: Geometry of Non-Cyclic Motion." *American Journal of Physics* 66: 431. https://doi.org/10.1119/1.18799

Gilmore, R. 1974. *Lie Algebra and Some Applications*. New York: Wiley.

Holstein, B. R. 1991. "Variations on the Aharonov-Bohm Effect." *American Journal of Physics* 59: 1080–1085. https://doi.org/10.1119/1.16617

Hong-Mo, C., and T. S. Tsun. 1993. *Some Elementary Gauge Theory Concepts*. Singapore: World Scientific.

Mukunda, N., and R. Simon. 1993. "Quantum Kinematics Approach to the Geometric Phase. I. General Formalism." *Annals of Physics (New York)* 228: 203. https://doi.org/10.1006/aphy.1993.1093

Pancharatnam, S. 1956. "Generalized Theory of Interference and Its Applications: Part I. Coherent Pencils." *Proceedings of the Indian Academy of Sciences A* 44: 247. https://doi.org/10.1007/BF03046050

Peshkin, M., and H. J. Lipkin. 1995. "Topology, Locality, and Aharonov-Bohm Effect with Neutrons." *Physical Review Letters* 74: 2847. https://doi.org/10.1103/PhysRevLett.74.2847

Shapere, A., and F. Wilczek, eds. 1989. *Geometric Phases in Physics*. Singapore: World Scientific.

Wilczek, F., and A. Zee. 1984. "Appearance of Gauge Structure in Simple Dynamical Systems." *Physical Review Letters* 52: 2111. https://doi.org/10.1103/PhysRevLett.52.2111

Zee, A. 1988. "Non-Abelian Gauge Structure in Nuclear Quadrupole Resonance." *Physical Review A* 38: 1. https://doi.org/10.1103/PhysRevA.38.1

CHAPTER 9

AMPLITUDE SQUEEZING IN SEMICONDUCTOR LASERS AND IN KERR INTERFEROMETER

It has been shown in various studies, by Yamamoto and his colleagues (Machida and Yamamoto 1988; Machida and Yamamoto 1989; Machida, Yamamoto, and Itaya 1987; Yamamoto, Machida, and Nilsson 1986; Yamamoto, Machida, and Richardson 1992) that the quantum noise of the amplitude of the EM field, in semiconductor lasers, can be reduced below the standard quantum limit (SQL), at the expense of increased quantum phase fluctuations. Other researchers have also reported similar results (e.g., Richardson and Shelby 1990). In the present work (Jann and Ben-Aryeh 1996), we follow the standard photoelectron detection theory described in Chapter 2. According to this theory, amplitude squeezing is defined by a Mandel Q-factor (Kimble, Dagenais, and Mandel 1977; Mandel and Wolf 1995), less than 1, where the Mandel Q-factor is given by:

$$Q = \frac{\langle n^2 \rangle - \langle n \rangle^2}{\langle n \rangle} - 1 \tag{9.1}$$

The interpretation of this definition is as follows: for coherent light $\langle n^2 \rangle - \langle n \rangle^2 = \langle n \rangle$ the Mandel Q-factor equals 0, while a negative Q-factor represents sub-Poisson photon statistics. The lower bound for the Q-factor is obtained for a number state where $Q = -1.$ An equivalent expression for the Mandel Q-factor can be written as $Q = \langle n \rangle [g^{(2)}(0) - 1]$, where $g^{(2)}(0) = \langle \hat{a}^\dagger \hat{a}^\dagger \hat{a} \hat{a} \rangle / \langle \hat{a}^\dagger \hat{a} \rangle^2$. According to photon statistics, a negative value of $[g^{(2)}(0) - 1]$ represents nonclassical light, so that the negative Mandel Q-factor serves as a signature of nonclassicality. It should be noted that our definition for amplitude squeezing is different from that used in previous studies (Machida and Yamamoto 1988; Machida and

Yamamoto 1989; Machida, Yamamoto, and Itaya 1987; Yamamoto, Machida, and Nilsson 1986; Yamamoto, Machida, and Richardson 1992).

9.1. Amplitude squeezing in semiconductor lasers using Langevin equations

In the present work, we follow the laser theory developed by Lax and Louisell (1967) employing a theoretical model that differs from that used in previous studies (Machida and Yamamoto 1988; Machida and Yamamoto 1989; Machida, Yamamoto, and Itaya 1987; Yamamoto, Machida, and Nilsson 1986; Yamamoto, Machida, and Richardson 1992), yet yields similar results. Conventional laser theories rely on adiabatic elimination of the transition dipole moment and the population inversion, which exhibit extensive decay times (Sargent, Scully, and Lamb 1978). After such eliminations, one obtains a quantum equation for the EM field, which, far above threshold, describes a coherent field. The semiconductor laser theory is based on the adiabatic elimination of only the transition dipole moment. Due to rapid fluctuations in the carrier density of semiconductor lasers (the analog of population inversion), the carrier density cannot be eliminated adiabatically.

We consider a single-mode resonator with annihilation and creation operators $\hat{a}$ and $\hat{a}^\dagger$, which is in resonance with the active laser medium. Denoting the photon density operator by $\hat{N} = \hat{a}^\dagger \hat{a}$, and the carrier density operator by $\hat{R}$, one gets the quantum mechanical Langevin equations (Lax and Louisell 1967):

$$\frac{d\hat{N}}{dt} = \hat{A}_N + \hat{F}_N \quad ; \quad \frac{d\hat{R}}{dt} = \hat{A}_R + \hat{F}_R, \tag{9.2}$$

where $\hat{A}_N$ and $\hat{A}_R$ are drift operators given by:

$$\begin{aligned} \hat{A}_N &= \beta\left(\hat{N}+1\right)\hat{R} + \gamma n_{th} - \gamma\hat{N} \quad ; \\ \hat{A}_R &= P - \Gamma\hat{R} - \beta\left(\hat{N}+1\right)\hat{R}. \end{aligned} \tag{9.3}$$

Here, $\hat{F}_N$ and $\hat{F}_R$ are the Langevin force operators for photon and carrier density fluctuations. Other operators are defined as follows: β, the average rate for spontaneous emission into the laser mode; γ, the overall loss rate of the cavity, where $1/\gamma$ is the coherence time of the laser; Γ, the carrier decay rate constant; P, the pump rate; and $n_{Th} = [\exp(\hbar\omega / kT) - 1]^{-1}$, the average number of thermal photons in the laser cavity.

Yamamoto and his colleagues (Machida and Yamamoto 1988; Machida and Yamamoto 1989; Machida, Yamamoto, and Itaya 1987; Yamamoto, Machida, and Nilsson 1986; Yamamoto, Machida, and Richardson 1992) showed that their lasers generate a number-phase squeezed state if the pump noise is eliminated. Following their experiments, they suggested that the oscillator produces an amplitude-squeezed state if the pump noise is suppressed below the SQL (Yamamoto, Machida, and Nilsson 1986). They developed various schemes for eliminating the pump noise fluctuations. In our theoretical model, *we eliminate the fluctuation of the parameter* P*, from our equations* (in agreement with pump noise suppression as described by Machida and Yamamoto 1988; Machida and Yamamoto 1989; Machida, Yamamoto, and Itaya 1987; Yamamoto, Machida, and Nilsson 1986; Yamamoto, Machida, and Richardson 1992), *and study the possibility that additional factors may also play a significant role in achieving results below the SQL.* Additionally, we assume that the experiments are done at very low temperatures so that the thermal noise can be neglected.

In the approximation of a large photon number, and when the laser oscillator is pumped well above threshold, we can apply quasi-linearization to expand $\hat{N}$ and $\hat{R}$ about the operating point:

$$\hat{N} = \bar{N} + \hat{n} \quad ; \quad \hat{R} = \bar{R} + \hat{r}. \tag{9.4}$$

Here, $\bar{N} = \langle \hat{N} \rangle$ and $\bar{R} = \langle \hat{R} \rangle$, $\hat{n}$ and $\hat{r}$ are slight deviations from these values, respectively. The drift terms of Eq. (9.4) are then expanded, and only linear *terms* in the deviations are retained. After using this procedure, Eq. (9.3) is reduced to the linear form:

$$\frac{d\hat{n}}{dt} = A\hat{r} - B\hat{n} + \hat{F}_n \quad ; \quad \frac{d\hat{r}}{dt} = -C\hat{r} - D\hat{n} + \hat{F}_r \quad , \tag{9.5}$$

where

$$A=\beta\bar{N} \quad ; \quad B=\gamma-\beta\bar{R} \quad ; \quad C=\Gamma+\beta\bar{N} \quad ; \quad D=\beta\bar{R} \, . \tag{9.6}$$

Here, we neglected the thermal photons and pump fluctuations.

Diffusion constants are calculated by the second moments of the noise sources:

$$\left\langle \hat{F}_{\alpha}(t)\hat{F}_{\beta}(u)\right\rangle = 2D_{\alpha\beta}\delta(t-u) \, . \tag{9.7}$$

Using conventional laser theory (Lax and Louisell 1967), the diffusion constants D_{nn} and D_{rr} are given by:

$$2D_{nn}=\beta\bar{N}\bar{R}+\gamma\bar{N} \quad ; \quad 2D_{rr}=\Gamma\bar{R}+\beta\bar{N}\bar{R} \, . \tag{9.8}$$

It has been shown by Lax and Louisell (1967) that there is another diffusion coefficient that describes a cross-correlation between the fluctuating forces in the corresponding reservoirs:

$$2D_{nr}=-\beta\bar{N}\bar{R} \, . \tag{9.9}$$

This diffusion constant arises because both reservoirs are coupled to the same dipole moment, i.e., the decay of the carrier is accompanied by photon generation, and photon absorption is accompanied by carrier excitation. Therefore, the reservoirs are negatively correlated with the dipole moment.

We used the following procedure to obtain the Mandel *Q*-factor for the present semiconductor laser: a) We transformed Eqs. (9.5-9.6) into Fourier space and solved them to obtain the solutions for $\hat{n}(\omega)$, and $\hat{r}(\omega)$; b) We assumed the relations:

$$\langle \hat{n}(\omega)\hat{n}^{\dagger}(\omega')\rangle = n_{\omega}^{2}\delta(\omega-\omega') \quad ; \quad \langle \hat{n}(\omega)\hat{n}(\omega')\rangle = 0 \, , \tag{9.10}$$

and

$$\langle F_{\mu}(\omega)F_{\nu}(\omega')\rangle = 2D_{\mu\nu}\delta(\omega-\omega') \, , \tag{9.11}$$

where the parameters $D_{\mu\nu}$ are given in Eqs. (9.8-9.9). Using these expressions, the parameter n_{ω}^{2} was calculated; c) Since

$$n_{\omega}{}^{2}=\int_{-\infty}^{\infty}\hat{n}(0)\hat{n}(\tau)\exp(-i\omega\tau)d\tau \quad , \tag{9.12}$$

the inverse transform with $\tau=0$ is given as:

$$\langle\hat{n}^{2}(0)\rangle=\int_{-\infty}^{\infty}n_{\omega}{}^{2}d\omega \quad . \tag{9.13}$$

The integral of Eq. (9.13) is evaluated by contour integration, closing the contour in the upper half of the complex plane and applying Jordan's lemma (Arfken 1985), yielding the result:

$$\langle\hat{n}^{2}(0)\rangle=\frac{A^{2}D_{rr}+2ACD_{rn}+(AD+BC+C^{2})D_{nn}}{(B+C)(AD+BC)} \quad , \tag{9.14}$$

where the parameters A,B,C,D were defined in Eq. (9.6), and the diffusion coefficients are given in Eqs. (9.8-9.9). The operating point is set to zero, and the following approximations are assumed: $\overline{N} \gg 1$ and $\beta\overline{R}=\gamma$.

The Mandel Q-parameter was calculated as a function of the average photon number $\overline{N}$, using typical parameter values for semiconductor lasers: $\Gamma=10^{5}\overline{N}\ \mathrm{sec}^{-1}$, $\beta=10^{5}\ \mathrm{sec}^{-1}$, and $\gamma=10^{11}\ \mathrm{sec}^{-1}$. We found that for $\overline{N}>2\times10^{6}$ the Mandel Q-factor becomes negative, i.e., falls below the SQL. This result depends critically on the parameter values. For example, by reducing the parameter D_{rn} to half of its value eliminates the negative Q-value, resulting in Poissonian statistics (see Fig. 1 in Jann and Ben-Aryeh 1996).

9.2. Amplitude squeezing in Kerr interferometer

We treated the effects of the beam splitters on a nonlinear Mach-Zehnder interferometer, where a Kerr medium is located on one of its arms (Ben-Aryeh and Zahler 1991a; Ben-Aryeh and Zahler 1991b). The calculations were carried out under very general conditions, resulting in expressions that are quite complex. We find that, to treat amplitude squeezing in a Kerr interferometer, it is more efficient to consider special conditions with special number-phase relations, following the analysis of Kitagawa and Yamamoto (1986).

9.3. Phase modulation in Kerr medium

The Hamiltonian of the single-mode electric field in Kerr medium can be written as:

$$H = H_0 + H_k \quad ; \quad H_0 = \hbar\omega_0 \hat{a}^\dagger \hat{a} = \hbar\omega_0 \hat{n},$$
$$H_K = \hbar\chi \left(\hat{a}^\dagger\right)^2 \hat{a}^2 = \hbar\chi \hat{n}(\hat{n}+1). \tag{9.15}$$

Here, the parameter χ is real and related to the nonlinear refractive index n_2 of Kerr medium:

$$\chi = \frac{\hbar\omega_0 n_2}{2c\varepsilon_0 n_0{}^2 A\tau}. \tag{9.16}$$

The travelling wave propagates at the speed $v = c/n_0$, where n_0 is the linear refractive index of Kerr medium, c is the speed of light in vacuum, ω_0 is the angular frequency, and τ is the coherence time, with a coherence length given by $l = \tau v$. The optical wave travelling through Kerr medium is assumed to be plane, polarized, and monochromatic. The quantization volume is denoted by $V = Al$, where A is the cross-sectional area of the optical beam. The intensity-dependent refractive index is expressed as:

$$\eta\left(|\mathrm{E}|^2\right) = n_0 + \frac{1}{2} n_2 |\mathrm{E}|^2, \tag{9.17}$$

where $E(z,t) = |E(z,t)| \exp(ikz - \omega_0 t)/2$, and $k = \omega_0 / v$ is the corresponding wavevector. This Hamiltonian is valid under the conditions of large detuning from the transition levels, no saturation, and no loss.

The state evolution during light propagation over a distance z, in the interaction picture, is given by:

$$|\Psi(z)\rangle = U_K(z)|\Psi_1\rangle = U_K(z)|\alpha_1\rangle, \tag{9.18}$$

where $U_K(z)$ is a unitary operator generated by the Hamiltonian H_K, and $|\Psi_1\rangle = |\alpha_1\rangle$ is the state at the entrance to Kerr medium. The state at the exit of Kerr medium of length L is:

$$|\psi(L)\rangle = U_K(L)|\alpha_1\rangle \ ; \ U_K(L) = \exp\left[\frac{i}{2}\gamma\hat{n}(\hat{n}-1)\right]; \gamma = \frac{2\chi L}{v} . \tag{9.19}$$

We assume that the initial state $|\psi_1\rangle$ is a coherent state $|\alpha_1\rangle$. Then, the density operator in the $Q(\alpha)$ representation (as described in Chapter 6) is given by:

$$\langle\alpha|\rho_K|\alpha\rangle = \langle\alpha|U_K(L)|\alpha_1\rangle\langle\alpha_1|U_K^\dagger(L)|\alpha\rangle . \tag{9.20}$$

By straightforward calculation of Eq. (9.20) (see the detailed calculations in Kitagawa and Yamamoto 1986), we obtain the 'quasi probability density' (*QPD*), for the EM field output from the Ker medium:

$$Q(\alpha,\alpha_1)_{Kerr} = exp\left[-\left(|\alpha|^2+|\alpha_1|^2\right)\right]\left|\sum_{n-0}^{\infty}\frac{(\alpha^*\alpha_1)^n}{n!}\exp\left[\frac{i}{2}\gamma n(n-1)\right]\right|^2 . \tag{9.21}$$

Numerical calculation of the *QPD* of the EM output from Kerr medium, based on Eq. (9.21), shows that it exhibits a crescent shape, wider in the middle than its ends, resembling the moon at its first and last quarters (Kitagawa and Yamamoto 1986). This is fundamentally different from the '*elliptical*' squeezing observed in ordinary squeezed states.

9.4. Reduction of photon number uncertainty by interference

We assume that the input in the Mach-Zehnder interferometer, before the first beam splitter BS_1 is a coherent state with $\alpha = \alpha_{in}$ in one input port, and the vacuum in the other input port. The states after the first beam splitter BS_1, are given by two coherent states:

$$|\Psi_1\rangle_{in} = |\alpha_1\rangle \ ; \ \alpha_1 = (1-R_1)^{1/2}\alpha_{in},$$

$$|\Psi_2\rangle_{in} = |\alpha_2\rangle \ ; \ \alpha_2 = (R_1)^{1/2}\alpha_{in}. \tag{9.22}$$

Here, R_1 (real) is the reflectivity of BS_1, while α_1 and α_2 are complex numbers. The state $|\alpha_1\rangle$ becomes the nonclassical state after propagation through Kerr medium, given by Eqs. (9.18-9.19) (including the *QPD* calculated in the $Q(\alpha)$ representation), while the state $|\alpha_2\rangle$ remains a coherent state $|\alpha_2 e^{i\theta}\rangle$.

The state $|\Psi_1\rangle_{out}$ after the second beam splitter, BS_2 is given by:

$$|\Psi_1\rangle_{out} = (R_2)^{1/2} U_K(L)|\alpha_1\rangle + (1-R_2)^{1/2}|\alpha_2 e^{i\theta}\rangle, \tag{9.23}$$

where R_2 (real) is the reflectivity of the second beam splitter BS_2, and $U_K(L)$ is defined in Eq. (9.19). We focus on the analysis of the state $|\Psi_1\rangle_{out}$, since under the condition that R_2 is nearly equal 1, $(1-R_2)^{1/2} \ll 1$, the output state $|\Psi_1\rangle_{out}$ becomes a well-defined nonclassical light state (unlike $|\Psi_2\rangle_{out}$). Under these conditions, it has been shown by Kitagawa and Yamamoto (1986) that $(1-R_2)^{1/2}|\alpha_2 e^{i\theta}\rangle = \xi$ is a nonclassical shift term, represented by the displacement operator $\hat{D}(\xi) = \exp(\xi a^{\dagger} - \xi^{*} a)$, so that under these conditions $|\Psi_1\rangle_{out}$ becomes:

$$|\Psi_1\rangle_{out} = \hat{D}(\xi)\hat{U}_K(L)|\alpha_1\rangle. \tag{9.24}$$

Here, $\hat{U}_K(L)|\alpha_1\rangle$ in Eq. (9.24) represents the transformation from a classical coherent state $|\alpha_1\rangle$ to a nonclassical state produced by Kerr medium. Its effect differs from that of ordinary squeezing, as it produces a ‘crescent’ shape rather than an ‘elliptical’ one. The operator $\hat{D}(\xi)$ displaces the center of the *QPD* of the present nonclassical state, by the shift ξ. Although $(1-R_2)^{1/2}$ appearing in Eq. (9.23) is quite small, it is multiplied by $|\alpha_2| \gg |\alpha_1|$, so the resulting displacement $|\xi|$ can become significant.

While the above analysis was carried out in the Schrödinger picture, we now use the Heisenberg picture to describe the annihilation operator $\hat{c}$ and the number operator $\hat{n}_c$ of the output mode in port 1 as:

$$\hat{c} = \hat{U}_K^\dagger(L)\hat{D}^\dagger(\xi)\hat{a}\hat{D}(\xi)\hat{U}_K \quad ; \quad \hat{n}_c = \hat{c}^\dagger\hat{c} \, . \tag{9.25}$$

Detailed calculations were presented in Kitagawa and Yamamoto (1986) for:

$$\sigma = (\Delta\hat{n}_c^2)/\langle\hat{n}_c\rangle \, , \tag{9.26}$$

as a function of various parameters. One may notice that Eq. (9.26) corresponds to the Mandel *Q*-factor plus 1, so that $\sigma < 1$ represent a sub-Poissonian distribution. The analytical results are rather complex, and numerical calculations demonstrated the nonclassical nature of these states.

9.5. Number-phase uncertainty relations and enhanced phase uncertainty

In Chapter 7, we extensively discussed phase operators and distributions. However, as shown by Kitagawa and Yamamoto (1986), when addressing these issues for a nonlinear interferometer with a Kerr medium, it is more effective to adopt the approach that Jackiw (1968) developed for the number-phase uncertainties, which proceeds as follows.

The boson annihilation and creation operators $\hat{a}$ and $\hat{a}^\dagger$ are decomposed into photon number $\hat{n}$ and phase factors $E_\pm$ as:

$$\hat{E}_- = (\hat{n}+1)^{-1/2}\hat{a} \quad ; \quad \hat{E}_+ = (\hat{n}+1)^{-1/2}\hat{a}^\dagger \, . \tag{9.27}$$

Well-defined operators are given by:

$$\hat{S} = \frac{1}{2i}\left(\hat{E}_- - \hat{E}_+\right) \quad ; \quad \hat{C} = \frac{1}{2}\left(\hat{E}_- + \hat{E}_+\right) \, , \tag{9.28}$$

With the corresponding commutation relations (CR):

$$\left[\hat{n},\hat{S}\right]=i\hat{C} \quad ; \quad \left[\hat{n},\hat{C}\right]=-i\hat{S} \tag{9.29}$$

The Heisenberg uncertainty relations follow from Eq. (9.29) as:

$$\langle\Delta\hat{n}^2\rangle\left\langle\Delta\hat{S}^2\right\rangle\geq\frac{1}{4}\left\langle\hat{C}\right\rangle^2\Rightarrow\frac{\langle\Delta\hat{n}^2\rangle\left\langle\Delta\hat{S}^2\right\rangle}{\left\langle\hat{C}\right\rangle^2}\geq\frac{1}{4}\ ;$$
$$\langle\Delta\hat{n}^2\rangle\left\langle\Delta\hat{C}^2\right\rangle\geq\frac{1}{4}\left\langle\hat{S}\right\rangle^2\Rightarrow\frac{\langle\Delta\hat{n}^2\rangle\left\langle\Delta\hat{C}^2\right\rangle}{\left\langle\hat{S}\right\rangle^2}\geq\frac{1}{4}. \tag{9.30}$$

The basic idea behind the number-phase relations (Kitagawa and Yamamoto 1986) is that the operators $E_\pm$ act on highly excited states, *approximately as unitary operators*, since these states are nearly orthogonal to the vacuum state $|0\rangle$. Therefore, a Hermitian phase operator Φ was adopted in the form:

$$E_\pm=\exp[\pm i\Phi] , \tag{9.31}$$

for highly coherent states (Haken 1970). In the Heisenberg picture, the operators $\hat{a},\ \hat{b}=\exp(i\gamma\hat{n})\hat{a}$ and $\hat{c}$ defined in Eq. (9.25), can be written as

$$\hat{a}=(\hat{n}_a+1)^{1/2}\exp\left[i\hat{\Phi}_a\right]\ ;\ \hat{b}=(\hat{n}_b+1)^{1/2}\exp\left[i\hat{\Phi}_b\right]\ ;$$
$$\hat{c}=(\hat{n}_c+1)^{1/2}\exp\left[i\hat{\Phi}_c\right]. \tag{9.32}$$

Here $\hat{a}$ is the annihilation operator before Kerr medium, $\hat{b}$ is the annihilation operator after Kerr medium, and $\hat{c}$ is the annihilation operator after the second beam splitter BS_2. Number and phase uncertainties were calculated for the nonlinear Kerr interferometer, adopting the definitions of Eq. (9.32), including the phase operators as formulated by Kitagawa and Yamamoto (1986).

Proposals for feedback quantum non-demolition measurements have been developed in various studies (e.g., Imoto, Haus, and Yamamoto 1985; Yamamoto, Imoto, and Machida 1986). Time-delay effects in such schemes were analyzed by Jann and Ben-Aryeh (1997).

References

Arfken, G. 1985. *Mathematical Methods for Physicists*. 3rd ed. Academic Press, p. 408.

Ben-Aryeh, Y., and M. Zahler. 1991a. "Beam Splitter Effects in a Nonlinear Kerr Medium." *Physica B* 175: 133. https://doi.org/10.1016/0921-4526(91)90702-G

Ben-Aryeh, Y., and M. Zahler. 1991b. "The Effect of the Beams-Splitter on the Quantum Detection Properties of the Nonlinear Mach-Zehnder Interferometer." *Optics Communications* 85: 136–42. https://doi.org/10.1016/0030-4018(91)90059-M

Haken, H. 1970. *Light and Matter*, Vol. XXV, in *Handbuch der Physik*. Berlin: Springer.

Imoto, N., H. A. Haus, and Y. Yamamoto. 1985. "Quantum Non Demolition Measurement of the Photon Number via the Optical Kerr Effect." *Physical Review A* 32: 2287. https://doi.org/10.1103/PhysRevA.32.2287

Jackiw, R. 1968. "Minimum Uncertainty Product, Number-Phase Uncertainty Product and Coherent States." *Journal of Mathematical Physics* 9: 339. https://doi.org/10.1063/1.1664585

Jann, A., and Y. Ben-Aryeh. 1996. "Quantum Noise Reduction in Semiconductor Lasers." *Journal of the Optical Society of America B* 13: 761. https://doi.org/10.1364/JOSAB.13.000761

Jann, A., and Y. Ben-Aryeh. 1997. "Time Delay Effects in Feedback Quantum Non Demolition Measurements." *Journal of the Optical Society of America B* 14: 11. https://doi.org/10.1364/JOSAB.14.000011

Kimble, H. J., M. Dagenais, and L. Mandel. 1977. "Photon Antibunching in Resonance Fluorescence." *Physical Review Letters* 39: 691. https://doi.org/10.1103/PhysRevLett.39.691

Kitagawa, M., and Y. Yamamoto. 1986. "Number-Phase Minimum Uncertainty in a Kerr Nonlinear Interferometer." *Physical Review A* 34: 3974. https://doi.org/10.1103/PhysRevA.34.3974

Lax, M., and W. Louisell. 1967. "Quantum Noise IX. Quantum Fokker-Planck Solution for Laser Noise." *IEEE Journal of Quantum Electronics* QE-3: 47–58. https://doi.org/10.1109/JQE.1967.1074446

Machida, S., and Y. Yamamoto. 1988. "Ultrabroadband Amplitude Squeezing in a Semiconductor Laser." *Physical Review Letters* 60: 792. https://doi.org/10.1103/PhysRevLett.60.792

Machida, S., and Y. Yamamoto. 1989. "Observation of Amplitude Squeezing from a Semiconductor Laser by Balanced Direct Detectors with a Delay Line." *Optics Letters* 14: 1045. https://doi.org/10.1364/OL.14.001045

Machida, S., Y. Yamamoto, and Y. Itaya. 1987. "Observation of Amplitude Squeezing in a Constant Current-Driven Semiconductor Laser." *Physical Review Letters* 58: 1000. https://doi.org/10.1103/PhysRevLett.58.1000

Mandel, L., and E. Wolf. 1995. *Optical Coherence and Quantum Optics*. Cambridge: Cambridge University Press. https://doi.org/10.1017/CBO9781139644105

Richardson, W. H., and R. M. Shelby. 1990. "Nonclassical Light from a Semiconductor Laser Operating at 4 K." *Physical Review Letters* 64: 400. https://doi.org/10.1103/PhysRevLett.64.400

Sargent, M., M. O. Scully, and W. E. Lamb. 1978. *Laser Physics*. New York: Best View Press.

Yamamoto, Y., S. Machida, and O. Nilsson. 1986. "Amplitude Squeezing in a Pump-Noise Suppressed Laser Oscillator." *Physical Review A* 34: 4025. https://doi.org/10.1103/PhysRevA.34.4025

Yamamoto, Y., S. Machida, and W. H. Richardson. 1992. "Photon-Number Squeezed States in Semiconductor Laser." *Science* 255: 1219. https://doi.org/10.1126/science.255.5049.1219

Yamamoto, Y., N. Imoto, and S. Machida. 1986. "Amplitude Squeezing in a Semiconductor Laser Using Quantum Non Demolition Measurement and Negative Feedback." *Physical Review A* 33: 3243.

PART II

SPECIAL SYSTEMS ANALYZED BY QUANTUM OPTICS METHODS

CHAPTER 10

INHIBITION OF ATOMIC DIPOLE COLLAPSES BY SQUEEZED LIGHT: PHOTONS AND ATOMS SQUEEZING

There is considerable interest in studying the interaction between atoms and nonclassical radiation. The inhibition or enhancement of atomic dipole decay due to a squeezed reservoir plays an important role in this field (Gardiner 1986; Gardiner and Savage 1984). The work of Gardiner and Savage (1984), using second-order perturbation theory, has served as a foundation for treating the quantum noise reduction in various optical systems. We found it of interest to investigate the analogous effects in pure quantum systems, e.g., in the Jaynes-Cummings model (JCM) with a one-mode squeezed vacuum state (Ben-Aryeh et al. 1992). Gea-Banacloche, Schlicher, and Zubairy (1988) showed that the emission spectrum of such systems is insensitive to the initial phase difference between the atomic dipole and the squeezed light. However, we proposed that, within a short time limit, the results of the JCM would resemble those of Gardiner's Markovian approximation. As expected, our results based on a pure Hamiltonian differ from those of Gardiner for longer times. In the following analysis, we analyze the inhibition of the dipole decay for a single atom in the environment of a one-mode squeezed vacuum state (Ben-Aryeh et al. 1992).

10.1. Inhibition of atomic dipole decay by a strong squeezed vacuum environment

The enhancement or suppression of atomic dipole decay is simplified under resonance conditions, in which the Hamiltonian in the interaction picture reduces to:

$$H_I = g\left(\sigma^+ a + \sigma^- a^\dagger\right), \tag{10.1}$$

where a and $a^\dagger$ are the annihilation and creation operators, respectively; σ^+, σ^-, and σ_z are the spin raising, lowering, and inversion operators, respectively; and g is the coupling constant. The propagator in the interaction picture, under resonance conditions, can be expanded as (Cohen, Mann, and Ben-Aryeh 1993):

$$U = \begin{bmatrix} \cos\left(gt\sqrt{aa^\dagger}\right) & -ia\dfrac{\sin\left(gt\sqrt{a^\dagger a}\right)}{\sqrt{a^\dagger a}} \\ -ia^\dagger\dfrac{\sin\left(gt\sqrt{aa^\dagger}\right)}{\sqrt{aa^\dagger}} & \cos\left(gt\sqrt{a^\dagger a}\right) \end{bmatrix}, \tag{10.2}$$

where the initial state in the present JCM is given by:

$$|\psi(0)\rangle = P(0)|a\rangle + Q(0)|b\rangle + \sum_{n=0}^{\infty} C_{2n}(0)|2n\rangle \quad . \tag{10.3}$$

Here, $P(0)$ and $Q(0)$ are the initial amplitudes of the atomic states, and $C_{2n}(0)$ is the initial amplitude for the pure one-mode squeezed vacuum state, given as (Cohen, Mann, and Ben-Aryeh 1993):

$$C_{2n}(0) = \frac{(-1)^n \exp(-in\phi)}{\sqrt{\cosh(r)}} \frac{\sqrt{(2n)!}}{2^n 2!} \left[\tanh(r)\right]^n , \tag{10.4}$$

where certain approximations to this equation were analyzed in Cohen, Mann, and Ben-Aryeh (1993).

The atom-field state at time t is given by (Cohen, Mann, and Ben-Aryeh 1993):

$$\begin{aligned}|\psi(t)\rangle &= U(t)|\psi(0)\rangle = Q(0)C(0)|0\rangle|b\rangle \\ &+\sum_{n=0}^{\infty} P(0)C_{2n}(0)\left\{\begin{matrix}\cos gt\sqrt{2n+1}\,|2n\rangle|a\rangle \\ -i\sin\left(gt\sqrt{2n+1}\right)|2n+1\rangle|b\rangle\end{matrix}\right\} \\ &+\sum_{n=0}^{\infty} Q(0)C_{2n}(0)\left\{\begin{matrix}\cos gt\sqrt{2n}\,|2n\rangle|b\rangle \\ -i\sin\left(gt\sqrt{2n}\right)|2n-1\rangle|a\rangle\end{matrix}\right\}.\end{aligned} \tag{10.5}$$

We calculate the complex dipole:

$$\rho_{ab} = \langle a|\sum_{n-0}^{\infty}\langle n|\psi(t)\rangle\langle\psi(t)|n\rangle|b\rangle, \tag{10.6}$$

so that by using Eq. (10.6), we get after straightforward calculations:

$$\begin{aligned}\rho_{ab} &= \sum_{n=0}^{\infty} Q^{*}(0)P(0)\left|C_{2n}(0)\right|^{2}\cos\left(gt\sqrt{2n+1}\right)\cos\left(gt\sqrt{2n}\right) \\ &+\sum_{n=0}^{\infty} Q(0)P^{*}(0)C_{2n+2}(0)C^{*}_{2n}\sin\left(gt\sqrt{2n+1}\right)\sin\left(gt\sqrt{2n+2}\right).\end{aligned} \tag{10.7}$$

We define θ by using the relation:

$$Q(0)P^{*}(0) = |Q(0||P(0)|\exp(-i\theta), \tag{10.8}$$

and the property:

$$C_{2n+2}(0)C^{*}_{2n}(0) = |C_{2n+2}(0)||C^{*}_{2n}(0)|(-1)\exp(-i\phi), \tag{10.9}$$

of the expansion coefficients of the squeezed vacuum state. Then, the real part of ρ_{ab} is given by:

$$\begin{aligned}\langle\sigma_x(t)\rangle &= \frac{1}{2}(\rho_{ab}+\rho_{ba}) = |Q(0)||P(0)| \\ &\times\sum_{n=0}^{\infty}\left[\begin{matrix}\cos\theta|C_{20}(0)|^{2}\cos\left(gt\sqrt{2n+1}\right)\cos\left(gt\sqrt{2n}\right) \\ -\cos(\theta+\phi)|C_{2n+2}(0)||C^{*}_{2n}|\times\sin\left(gt\sqrt{2n+1}\right)\sin\left(gt\sqrt{2n+2}\right)\end{matrix}\right],\end{aligned} \tag{10.10}$$

and the imaginary part is given by:

$$\langle\sigma_y(t)\rangle = \frac{1}{2}(\rho_{ab} - \rho_{ba}) = |Q(0)||P(0)|$$
$$\times\sum_{n-0}^{\infty}\left[\begin{array}{l}\sin\theta|C_{20}(0)|^2\cos\left(gt\sqrt{2n+1}\right)\cos\left(gt\sqrt{2n}\right)\\ +\sin(\theta+\phi)|C_{2n+2}(0)||C^{\square}{}_{2n}|\sin\left(gt\sqrt{2n+1}\right)\sin\left(gt\sqrt{2n+2}\right)\end{array}\right]. \quad (10.11)$$

To simplify the results for the components of the complex dipole, we assume $\phi = 0$, and use the following approximations for strong squeezing:

$$|C_{2n}||C_{2n+2}| \approx |C_{2n}|^2 \quad ; \quad \sqrt{2(n+1)} \approx \sqrt{2n} \quad . \quad (10.12)$$

Then, by using all the above approximations, we get (Ben-Aryeh et al. 1992):

$$\langle\sigma_x(t)\rangle \approx \langle\sigma_x(0)\rangle\sum_{n=0}^{\infty}|C_{2n}|^2\cos\left(2gt\sqrt{2n+1}\right) ;$$
$$\langle\sigma_x(0)\rangle = |P(0)Q(0)|\cos\theta. \quad (10.13)$$

$$\langle\sigma_y(t)\rangle \approx \langle\sigma_y(0)\rangle\sum_{n=0}^{\infty}|C_{2n}|^2 ;$$
$$\langle\sigma_y(0)\rangle = |P(0)Q(0)|\sin(\theta). \quad (10.14)$$

We find that for short times $\langle\sigma_x(t)\rangle$, decays due to the interference between Rabi oscillations with different frequencies are proportional to n (represented by the term $\cos(2gt\sqrt{2n+1})$). By using asymptotic expansions for strong squeezing, this decay was found to be proportional to $\exp[-(gt)^2/\varepsilon]$, where $\varepsilon = 2\exp(-2r)$ (Schleich, Walls, and Wheeler 1988). In comparison with this decay, we find that $\langle\sigma_y(t)\rangle$ does not decay. This difference between the two components of the complex dipole is similar to that found in Gardiner's analysis (Gardiner and Savage 1984), using second-order perturbation theory.

10.2. Squeezing in a one-mode JCM

The one-mode JCM, in the rotating wave approximation, was given in Eq. (10.1). The propagator is given as (Cohen, Ben-Aryeh, and Mann 1994):

$$U = \begin{bmatrix} \cos(\mu) & -i\dfrac{\sin(\mu)}{\sqrt{aa^\dagger}}a \\ -i\dfrac{\sin(\nu)}{\sqrt{a^\dagger a}}a^\dagger & \cos(\nu) \end{bmatrix}$$

$$\mu = gt\sqrt{aa^\dagger} \quad ; \quad \nu = gt\sqrt{a^\dagger a} \tag{10.15}$$

where a and $a^\dagger$ are the annihilation and creation operators, respectively; σ^+, σ^-, and σ_z are the spin raising, lowering, and inversion operators, respectively; and g is the coupling constant (notice the small difference between the definition of *U* in Eq. (10.2) and in Eq. (10.15)).

It can be shown that if $\langle a \rangle = 0$, squeezing is obtained under the condition (Cohen, Ben-Aryeh, and Mann 1994):

$$\left|\langle a^2 \rangle\right| > \langle a^\dagger a \rangle . \tag{10.16}$$

Numerical calculations were performed (Cohen, Ben-Aryeh, and Mann 1994) to calculate the squeezing of the EM field in one-mode JCM, where the squeezing was defined (under the condition $\langle a \rangle = 0$) as:

$$D = \left|\langle a^2 \rangle\right| - \langle a^\dagger a \rangle , \tag{10.17}$$

so that for $D > 0$, the EM field is squeezed.

There are some important conclusions that can be drawn from the numerical calculations:

a) Better squeezing is obtained when the atom is initially in the lower state $|\psi_{at}\rangle = \begin{pmatrix} 0 \\ 1 \end{pmatrix}$. When the atom is initially in the upper state

$|\psi_{at}\rangle = \begin{pmatrix} 1 \\ 0 \end{pmatrix}$, its de-excitation emits an additional photon, which reduces the squeezing effect.

b) Squeezing is destroyed more rapidly for large values of the squeezing parameter r, since a single atom cannot support the presence of many squeezed photons.

10.3. Transfer of correlations from two-mode squeezed vacuum states of radiation to two atoms in a two-mode JCM

Yuen (1976) was the first to introduce the two-mode squeezed states, referred to as the "two-photon coherent states", and many authors have since studied their special features. A particularly interesting problem concerns how squeezed states behave when they interact with atoms. Experimentally, this leads to some difficulties, as all modes interacting with atoms should be "squeezed". In the work by Ficek (1991), the interaction of two-level atoms with broadband squeezed states was discussed; however, atomic correlations were not addressed. The inversion of an atom in the JCM, with a field initially in a one-mode squeezed vacuum, was investigated by Milburn (1984). Wang and Vyas (1992) studied the inversion of two two-level atoms interacting with a single mode squeezed state in JCM. Joshi and Puri (1990) investigated a two-mode squeezed state interacting with a single atom. Additional investigations of squeezing in JCM can be found in the work of Meystre and Zubairy (1982).

In the analysis described by Cohen, Ben-Aryeh, and Mann (1994), it was shown that the correlations between the two modes of a squeezed state can be transferred into correlations between two atoms interacting with this field. We considered a cavity supporting two distinct modes with frequencies ω_1 and ω_2, respectively. In addition, two different two-level atoms are placed inside the cavity. The first atom, which we denote by 1, is exactly resonant with the mode ω_1, and atom 2 is exactly resonant with the mode ω_2. The Hamiltonian describing this model is the sum of two JCM Hamiltonians:

$$H_1 = \hbar\omega_1 a_1^\dagger a_1 + \frac{\hbar\omega_1}{2}\sigma_1^z + \hbar g_1\left(a_1\sigma_1^+ + a_1^\dagger\sigma_1^-\right), \qquad (10.18)$$

and similarly for the second atom H_2, with index 2 replacing index 1. The time evolution of each system is determined by the Hamiltonian given in Eq. (10.18) with the appropriate index. The total propagator is the direct product of the two single-mode propagators:

$$U = U_1 \otimes U_2 . \tag{10.19}$$

We are interested in the evaluation of the products of operators of the two systems:

$$\hat{O}(t) = \hat{O}_1(t) \otimes \hat{O}_2(t) . \tag{10.20}$$

Since the two systems do not interact directly with each other

$$\hat{O}_1(t) = U_1^\dagger(t)\hat{O}_1(0)U_1(t) , \tag{10.21}$$

and similarly for the $O_2(t)$, each operator evolves as a function of time like in one-mode JCM.

The initial state of the entire system is given as:

$$|\psi_{init}\rangle = |\psi_1\rangle \otimes |\psi_2\rangle \otimes |\psi_f\rangle , \tag{10.22}$$

where $|\psi_1\rangle$ and $|\psi_2\rangle$ are the initial states of the two atoms, and $|\psi_f\rangle$ is the initial EM field state, given by:

$$|\psi_f\rangle = \frac{1}{\cosh(r)} \sum_n (\tanh r)^n \exp(in\phi)|n\rangle_1 |n\rangle_2 , \tag{10.23}$$

where r and ϕ are the squeezing parameters. Combining Eqs. (10.21) and (10.22), we get for the expectation value of the operator $\hat{O}(t)$ defined in Eq. (10.20), as:

$$\langle O(t)\rangle = \langle \psi_f | \langle \psi_1 | \hat{O}_1(t) | \psi_1 \rangle \langle \psi_2 | \hat{O}_2(t) | \psi_2 \rangle | \psi_f \rangle . \tag{10.24}$$

We introduce the shorthand notations:

$$\langle O_1(t)\rangle_{at} = \langle \psi_1 | O_1(t) | \psi_1 \rangle \;\; ; \;\; \langle O_2(t)\rangle_{at} = \langle \psi_2 | O_2(t) | \psi_2 \rangle \quad (10.25)$$

It should be noted that $\langle O_1(t)\rangle_{at}$ and $\langle O_2(t)\rangle_{at}$ are still operators, acting on the EM field.

Under the conditions $\langle a_1 \rangle = 0$, and $\langle a_2 \rangle = 0$ (Meystre and Zubairy 1982), we define a criterion for squeezing:

$$2|\langle a_1 a_2 \rangle| > \langle a_1^\dagger a_1 \rangle + \langle a_2^\dagger a_2 \rangle \quad (10.26)$$

This criterion can be calculated numerically using the framework described above. For example:

$$\langle a_1 a_2(t)\rangle = \langle \psi_f | \langle \psi_1 | a_1(t) | \psi_1 \rangle \langle \psi_2 | a_2(t) | \psi_2 \rangle | \psi_f \rangle \quad (10.27)$$

We calculated numerically the squeezing, defined (under the conditions $\langle a_1 \rangle = 0$. $\langle a_2 \rangle = 0$) by:

$$D = |\langle a_1 a_2 \rangle| - \frac{1}{2}\left(\langle a_1^\dagger a_1 \rangle + \langle a_2^\dagger a_2 \rangle\right) \quad (10.28)$$

From these numerical calculations, we draw conclusions similar to those found in the one-mode case.

1. Squeezing is improved when both atoms are in the ground state and when the two coupling constants are equal.
2. The decay of squeezing is faster for larger values of the squeezing parameter r.

10.4. Correlations of the atoms

Correlations between the two atoms were obtained through numerical calculations. Particularly interesting results were found for the *anomalous correlation functions* C_{++}. We define the atomic ground state as $\begin{pmatrix} 0 \\ 1 \end{pmatrix}$

(upper atomic state as $\begin{pmatrix}1\\0\end{pmatrix}$), and the dipole σ^+ operator as $\begin{pmatrix}0&1\\0&0\end{pmatrix}$ (dipole σ^- operator as $\begin{pmatrix}0&0\\1&0\end{pmatrix}$).

The *anomalous correlation function* C_{++} is then defined as:

$$C_{++}=\langle\psi_f|\langle\sigma_1^+(t)\sigma_2^+(t)\rangle|\psi_f\rangle=\langle\psi_f|\langle\sigma_1^+(t)\rangle\langle\sigma_2^+(t)\rangle|\psi_f\rangle;$$
$$\langle\sigma_1^+(t)\rangle=(0\quad 1)U^\dagger(t)\begin{pmatrix}0&1\\0&0\end{pmatrix}U(t)\begin{pmatrix}0\\1\end{pmatrix}=\sin\left(g_1t\sqrt{n+1}\right)\cos\left(g_1t\sqrt{n}\right),\tag{10.29}$$

where the matrix $U(t)$ is defined by Eq. (10.15), and $|\psi_f\rangle$ is defined in Eq. (10.23). After somewhat lengthy calculations, we get:

$$C_{++}=\langle\psi_f|\langle\sigma_1^+(t)\sigma_2^+(t)\rangle|\psi_f\rangle=-\frac{\exp(-i\phi)}{\cosh^2 r}$$
$$\sum_n(\tanh r)^{2n+1}\sin\left(g_1t\sqrt{n+1}\right)\cos\left(g_1t\sqrt{n}\right)\sin\left(g_2t\sqrt{n+1}\right)\cos\left(g_2t\sqrt{n}\right),\tag{10.30}$$

and similarly,

$$C_{--}=\langle\psi_f|\langle\sigma_1^-(t)\sigma_2^-(t)\rangle|\psi_f\rangle=C^{*}_{++}.\tag{10.31}$$

For the *normal correlation function*, we get:

$$C_{+-}=\langle\psi_f|\langle\sigma_1^+(t)\sigma_2^-(t)\rangle|\psi_f\rangle=0.\tag{10.32}$$

This is a general result, as the operators must create or annihilate the same number of photons to yield a nonzero expectation value. Based on the numerical calculations by Cohen, Ben-Aryeh, and Mann (1994), we find the following notable results:

1. For small squeezing, e.g., r = 0.5, with an average photon number of 0.27 for each mode, C_{++} has medium intensity, is positive, and

exhibits oscillatory behavior. The maxima of C_{++} corresponds to minima in the squeezing of light, and vice versa.

2. For strong squeezing, e.g., $r = 2$, with an average photon number of 13 in each mode, C_{++} has strong intensity and is approximately constant, while the squeezing of the light is minimal. These properties are consistent with the above interpretations.

10.5. Phase variance of squeezed states

Susskind and Glogower (SG) (Loudon 1983; Susskind and Glogower 1964) defined, in the Hilbert space of the harmonic oscillator, the following raising and lowering operators:

$$\exp\left(-i\hat{\phi}\right) \equiv \hat{E}_+ = \sum_{n=0}^{\infty} |n+1\rangle\langle n| \; ; \; \exp\left(i\hat{\phi}\right) \equiv \hat{E}_- = \sum_{n=0}^{\infty} |n\rangle\langle n+1| \quad . \tag{10.33}$$

Here, $|n\rangle \; (n = 0,1,2,\cdots,\infty)$ denotes the number states. Since the number states are bounded from below by the vacuum state, the SG operators are not unitary (Carruthers and Nieto 1968):

$$\hat{E}_-\hat{E}_+ = 1 \quad ; \quad \hat{E}_+\hat{E}_- = 1 - |0\rangle\langle 0| \; . \tag{10.34}$$

This non-unitarity leads to difficulties in the quantum description of the optical phase. This problem is solved either by using a finite-dimensional Hilbert space (Barnett and Pegg 1989; Pegg and Barnett 1989), or by applying the anti-normal ordering of the SG operators (Brif and Ben-Aryeh 1994; Luks and Perinova 1991). In Chapter 7, we briefly discussed general issues related to the definition of phase operators. Here, we focus on how phase operators and their variance can be defined for squeezed states. We find that all the above formalism becomes ill-defined for squeezed states (Cohen, Ben-Aryeh, and Mann 1992). For example, the expectation values of the operators E_+ and E_- vanish for the squeezed vacuum state.

Since squeezed states are related to two-photon processes, it is possible to introduce the non-unitary SG one-mode squeezed vacuum operator:

$$\left[\exp(i\hat{\phi})\right]^2 = \sum_{n=0}^{\infty} |n\rangle\langle n+2| \quad . \tag{10.35}$$

It was suggested by Cohen, Ben-Aryeh, and Mann (1992) to define the phase variance of the squeezed vacuum state as:

$$V(2\phi) = 1 - \left|\exp\left(i\hat{\phi}\right)\right|^2 = 1 - \left|\sum_{n-0}^{\infty} C^{*}_{n} C_{n+2}\right|^2 \quad . \tag{10.36}$$

Numerical calculations have shown the variance $V(2\phi)$ for one-mode squeezed vacuum, changes from the value 1 at $r = 0$ to $V(2\phi) = 0$ for $r \gg 1$, which seems to be quite reasonable.

A more general approach to the quantum phase problem is to adopt the method of positive operator-valued measure (POVM), as analyzed by Barak and Ben-Aryeh (2005). The phase states $|\theta\rangle$ are defined as (Carruthers and Nieto 1968):

$$|\theta\rangle = \frac{1}{\sqrt{2\pi}} \sum_{n=0}^{\infty} e^{in\theta} |n\rangle \quad . \tag{10.37}$$

These states are not orthogonal (Gea-Banacloche, Schlicher, and Zubairy 1988), due to the one-sided nature of the number spectrum, but they satisfy the identity resolution:

$$\int_0^{2\pi} d\theta |\theta\rangle\langle\theta| = \sum_{m,n=0}^{\infty} |m\rangle\langle n| \frac{1}{2\pi} \int_0^{2\pi} d\theta \exp\left[i(m-n)\theta\right] = \sum_{m,n=0}^{\infty} |m\rangle\langle n| \delta_{mn} = 1. \tag{10.38}$$

There are different operational methods for measuring phase distributions in one- and two-mode EM fields, which are based on phase states. As the phase states are not orthogonal, they are treated as non-orthogonal POVM operators. In Barak and Ben-Aryeh (2005), POVM phase distributions for one- and two-mode EM waves were studied. The key idea is that such distributions can be related to the density matrix, which is obtained from quadrature probability distribution measurements. For the one-mode case, the quantum phase distribution is given as:

$$P(\theta) = Tr\{\hat{\rho}|\theta\rangle\langle\theta|\} = \sum_{m.n=0}^{\infty} \rho_{mn} \exp[i(m-n)\theta] \quad . \tag{10.39}$$

The wavefunction for a one-mode EM field can be given as $|\psi\rangle = \Sigma_{n=0}^{\infty} C_n |n\rangle$ where C_n is the amplitude for the photon number state $|n\rangle$, corresponding to n photons. The quantum phase distribution for this state is then given by:

$$P(\theta) = (2\pi)^{-1} \sum_{m,n=0}^{\infty} C_m^{*} C_n \exp[i(m-n)\theta] \quad . \tag{10.40}$$

It may be noted that the special definition given in Eq. (10.36) accounts only for two-photon transitions $m-n=\pm 2$, while Eq. (10.40) represents the full phase distribution.

References

Barak, R., and Y. Ben-Aryeh. 2005. "Non-Orthogonal Positive Operator Valued Measure Phase-Distributions of One- and Two-Mode Electromagnetic Fields." *Journal of Optics B: Quantum and Semiclassical Optics* 7: 123. https://doi.org/10.1088/1464-4266/7/5/001

Barnett, S. M., and D. T. Pegg. 1989. "On the Hermitian Optical Phase Operator." *Journal of Modern Optics* 36: 7. https://doi.org/10.1080/09500348914550021

Ben-Aryeh, Y., C. A. Miller, H. Risken, and W. Schleich. 1992. "Inhibition of Atomic Dipole Collapses by Squeezed Light: A Jaynes-Cummings Model Treatment." *Optics Communications* 90: 259–64. https://doi.org/10.1016/0030-4018(92)90272-S

Brif, C., and Y. Ben-Aryeh. 1994. "Anti-Normal Ordering of Susskind-Glogower Quantum Phase." *Physical Review A* 50: 2727–37. https://doi.org/10.1103/PhysRevA.50.2727

Carruthers, P., and M. M. Nieto. 1968. "Phase and Angle Variables in Quantum Mechanics." *Reviews of Modern Physics* 40: 411. https://doi.org/10.1103/RevModPhys.40.411

Cohen, D., A. Mann, and Y. Ben-Aryeh. 1993. "Phase Effect for Atomic Dipoles in a Jaynes-Cummings Model with Various Initial Electromagnetic Fields." *Optics Communications* 99: 123–34. https://doi.org/10.1016/0030-4018(93)90715-H

Cohen, D., Y. Ben-Aryeh, and A. Mann. 1994. "Transfer of Correlations from Squeezed States of Radiation to Atoms in a Generalized Jaynes-Cummings Model." *Physical Review A* 49: 2040. https://doi.org/10.1103/PhysRevA.49.2040

Cohen, D., Y. Ben-Aryeh, and A. Mann. 1992. "Phase Variance of Squeezed States." *Optics Communications* 94: 227–30. https://doi.org/10.1016/0030-4018(92)90018-M

Ficek, Z. 1991. "Effect of a Broadband Squeezed Vacuum on Two-Atom Spontaneous Emission." *Optics Communications* 82: 130–34. https://doi.org/10.1016/0030-4018(91)90204-Q

Gardiner, C. W. 1986. "Inhibition of Atomic Phase Decays by Squeezed Light: A Direct Effect of Squeezing." *Physical Review Letters* 56: 1917. https://doi.org/10.1103/PhysRevLett.56.1917

Gardiner, C. W., and C. M. Savage. 1984. "Multimode Quantum Theory of a Degenerate Parametric Amplifier in a Cavity." *Optics Communications* 50: 173–76. https://doi.org/10.1016/0030-4018(84)90342-0

Gea-Banacloche, J., R. R. Schlicher, and M. S. Zubairy. 1988. "Emission Spectra for an Atom in a Cavity in the Presence of a Squeezed Vacuum." *Physical Review A* 38: 3514. https://doi.org/10.1103/PhysRevA.38.3514

Joshi, A., and R. R. Puri. 1990. "Characteristics of Rabi Oscillations in the Two-Mode Squeezed State of the Field." *Physical Review A* 42: 4336. https://doi.org/10.1103/PhysRevA.42.4336

Loudon, R. 1983. *The Quantum Theory of Light*. Oxford: Oxford University Press.

Luks, A., and V. Perinova. 1991. "Extended Number State Basis and Number-Phase Intelligent States of Light Field. I. Mapping and Operator Ordering Approach to a Quantum Phase Problem." *Czechoslovak Journal of Physics* 41: 1205–20. https://doi.org/10.1007/BF01613540

Meystre, P., and M. S. Zubairy. 1982. "Squeezed States in the Jaynes-Cummings Model." *Physics Letters A* 89: 390–94. https://doi.org/10.1016/0375-9601(82)90330-9

Milburn, G. J. 1984. "Interaction of a Two-Level Atom with Squeezed Light." *Optica Acta* 31: 671–77. https://doi.org/10.1080/10715769900301231

Pegg, D. T., and S. M. Barnett. 1989. "Phase Properties of the Quantized Single Mode Electromagnetic Field." *Physical Review A* 39: 1665–75. https://doi.org/10.1103/PhysRevA.39.1665

Schleich, W., D. F. Walls, and J. A. Wheeler. 1988. "Area of Overlap and Interference in Phase Space Versus Wigner Pseudoprobabilities." *Physical Review A* 38: 1177. https://doi.org/10.1103/PhysRevA.38.1177

Susskind, L., and J. Glogower. 1964. "Quantum Mechanical Phase and Time Operator." *Physics* 1: 49–61. https://doi.org/10.1103/PhysicsPhysiqueFizika.1.49

Wang, C., and R. Vyas. 1992. "Two-Photon Interaction of a Phased Atom with Squeezed Vacuum State." *Optics Communications* 91: 347–53. https://doi.org/10.1016/0030-4018(92)90358-X

Yuen, H. P. 1976. "Two-Photon Coherent States of the Radiation Field." *Physical Review A* 13: 2226–43. https://doi.org/10.1103/PhysRevA.13.2226

CHAPTER 11

INTRINSIC BISTABILITY OF HIGH-DENSITY TWO-LEVEL SYSTEMS AND OF EXCITED ATOMS

11.1. Introduction

Mirrorless or intrinsic optical bistability (IOB) has been investigated in several theoretical studies (e.g., Bowden and Sung 1979; Hopf, Bowden, and Louisell 1984). Semi-classical analyses for two-level systems have shown that local or Lorentz fields can introduce IOB (Bowden and Dowling 1993). In this chapter, we follow the quantum mechanical analysis presented by Ben-Aryeh, Bowden, and England (1986), as it is important to follow the various approximations in deriving IOB. While local field effects in linear systems are well known (Bowden and Dowling 1993), we treat here the non-linear local field effects.

The Heisenberg equations of motion for our system are obtained in the bad-cavity limit by adiabatically eliminating the variables associated with the field modes. We eliminate the rapid time and spatial dependence of the dipole operators by substituting:

$$\sigma_{+0}{}^{i} = \sigma_{+}{}^{i} \exp\left[-i\left(\omega_0 t - \vec{k}_0 \cdot \vec{r}_i\right)\right]. \tag{11.1}$$

This yields the following equations of motion:

$$\frac{d\sigma_z{}^{i}(t)}{dt} = -\beta_1\left[\sigma_z(t)+1\right] - \left|\sum_{j=1}^{N} \begin{array}{l} B^{\square}(i,j)\,\sigma_{+0}{}^{i}(t)\,\sigma_{-0}{}^{i}\left(t-\dfrac{r_{ij}}{c}\right) \\ \exp\left[i\vec{k}_0\cdot\left(\vec{r}_i-\vec{r}_j\right)+\sigma_{+0}{}^{i}(t)\left[\omega_R+2f^{+}(\vec{r}_i,t)\right]\right] \end{array}\right|. \tag{11.2}$$

$$\frac{d\sigma_{+o}{}^{i}(t)}{dt}=\left(i\omega-\beta_{2}\right)\sigma_{+o}{}^{i}(t)+$$

$$\left(\begin{array}{l}\sum_{j=1}^{N}B(i,j)\sigma_{+o}{}^{i}\left(t-\frac{r_{ij}}{c}\right)\sigma_{z}{}^{j}(t)\\ \exp\left[-i\vec{k}_{0}\cdot\left(\vec{r}_{i}-\vec{r}_{j}\right)\right]+\sigma_{z}{}^{i}\left[\frac{\omega_{R}}{2}+f^{-}\left(\vec{r}_{i},t\right)\right]\end{array}\right). \tag{11.3}$$

We use the standard operators related to the SU(2) algebra for a single atom: σ_z^i represents the population inversion operator, and $\sigma_\pm^i$ are the raising and lowering operators, for an atom i located at coordinates r_i. ω_R denotes the Rabi rate of the coherent field, while ω_0 and k_0 are the frequency and wavevector of the applied field, respectively. The summation on j indicates atoms with $j \neq i$ and $f^+(r_i,t)$, $f^-(r_i,t)$ are Langevin operators that stem from the vacuum contributions. These operators are δ correlated,

$$\left\langle f^{+}(r_{i},t),f^{-}(r_{i},t\right\rangle \propto \delta\left(t-t'\right). \tag{11.4}$$

The constants β_1 and β_2 may be considered as empirical parameters. In a fully quantum mechanical model, $\beta_1 = 2\beta$, and $\beta_2 = \beta$ where $\beta = 2\left|P\right|^2 k^3 / 3\hbar$ is defined as half the spontaneous decay rate, and P is the dipole matrix element.

In deriving Eqs. (11.2) and (11.3), we sum the interaction of the atom i with many other atoms, described by the summation over j $(j \neq i)$. We assume a model based upon many-body interactions in which each atom is influenced by the ensemble average of all other atoms. Such an assumption is justified when the number of atoms within a wavelength-scale volume is large ($\gg 1$). The equations are applied either under steady state conditions or for small fluctuations near the steady state. Under such conditions, any correlations due to initial conditions are destroyed by dephasing mechanisms, and the characteristic time of decorrelation can be related to the homogeneous broadening. As a result, we can justify the factorization of the products of dipole operators among different atoms. With this approximation, Eqs. (11.2) and (11.3) are transformed to:

$$\frac{d\left\langle \sigma_z^{(i)}(t)\right\rangle}{dt} = -\beta_1\left[\left\langle \sigma_z^{\,i}(t)\right\rangle + 1\right] - \left\{\left\{\sum_{j=1}^{N} \begin{matrix} B^{□}(i,j)\left\langle \sigma_{+0}^{\,i}(t)\right\rangle\left\langle \sigma_{-0}^{\,i}\right\rangle\left(t-\frac{r_{ij}}{c}\right) \\ \exp\left[i\vec{k}_0\cdot\left(\vec{r}_i-\vec{r}_j\right)\right] + \omega_R\sigma_{+0}^{\,i}(t) + C.C. \end{matrix}\right\}\right\}, \tag{11.5}$$

$$\frac{d\left\langle \sigma_{+o}^{\,i}(t)\right\rangle}{dt} = \left(i\Delta-\beta_2\right)\left\langle \sigma_{+o}^{\,i}(t)\right\rangle + \begin{pmatrix} \sum_{j=1}^{N} B(i,j)\left\langle \sigma_{+o}^{\,i}\left(t-\frac{r_{ij}}{c}\right)\right\rangle\left\langle \sigma_z^{\,j}(t)\right\rangle \\ \exp\left[-i\vec{k}_0\cdot\left(\vec{r}_i-\vec{r}_j\right)\right] + \omega^{□}_R\left\langle \sigma_z^{\,i}\right\rangle \end{pmatrix}. \tag{11.6}$$

Here, $\Delta = \omega - \omega_0$ is the deviation of the applied field frequency from resonance. We neglect the Langevin terms, as their expectation values vanish.

The reaction field is defined as:

$$\Theta\left(r_i\right) = \sum_{j=1}^{N} B(i,j)\exp\left[-i\vec{k}_0\cdot\left(\vec{r}_i-\vec{r}_j\right)\right]\sigma_{+0}^{\,j}\left(t-r_{ij}/c\right). \tag{11.7}$$

The cooperative terms in Eqs. (11.5) and (11.6) represent the effect of the reaction field. This field produces a first-order phase transition, where two phases of high and low excitations coexist along the direction of propagation through the material (Ben-Aryeh, Bowden, and England 1987). We develop the theory for two special cases: (1) A thin sample of two-level atoms, with a thickness smaller than a resonance wavelength λ. (2) A long sample of two-level atoms, with dimensions that are much larger relative to the wavelength.

11.2. Thin sample

For coherent radiation impinging on a thin film of two-level atoms, with a large surface area and a thickness smaller than the wavelength λ, we neglect the retardation time r_{ij}/c and apply the approximation:

$$\exp\left[i\vec{k}_0 \cdot \left(\vec{r}_i - \vec{r}_j\right)\right] = 1 . \tag{11.8}$$

Under this approximation and by using mean values for the expectation values of the atomic operators at the same time, Eqs. (11.5) and (11.6) are transformed into:

$$\frac{d\langle\sigma_z\rangle}{dt} = -\beta_1\left[\langle\sigma_z\rangle + 1\right] - 4\operatorname{Re}\Gamma\langle\sigma_{+0}\rangle\langle\sigma_{-0}\rangle + \omega_R\langle\sigma_{+0}(t)\rangle + \omega^*_R\langle\sigma_{-0}\rangle, \tag{11.9}$$

$$\frac{d\langle\sigma_{+o}\rangle}{dt} = (i\Delta - \beta_2)\langle\sigma_{+o}\rangle + \Gamma\langle\sigma_{+o}\rangle\langle\sigma_z\rangle + \omega^*_R\langle\sigma_z\rangle . \tag{11.10}$$

Here, we use the relation:

$$\Gamma = \sum_{j=1}^{N} B(i,j) . \tag{11.11}$$

In Ben-Aryeh, Bowden, and England (1986, Appendix A), Γ was calculated as a function of the position of an atom i in the thin sample. In these calculations, a volume $(4/3)\pi\varepsilon^3 = 1/n$ surrounding atom $\vec{r}_i$ was excluded from the integration range, where n is the atom density, in order to exclude the self-field of an atom i. The general calculations and resulting expressions are quite complex. However, in the limit $kd \to 0,\ k\varepsilon \to 0$, we get the approximation:

$$\Gamma = -i\frac{3\beta n\lambda^3}{8\pi^2} , \tag{11.12}$$

where $\beta = \beta_2$. Here, we neglected a small logarithmic dependence on the location of the atom within the thin sample. In more general calculations, we find that the real part of Γ is small relative to the imaginary part, and this ratio decreases with increasing atomic density. While the imaginary part of Γ leads to a renormalization of the frequency, the real part of Γ, although positive, may be related to a minor change in the semi-classical equations of motion.

Under the steady state condition, Eq. (11.10) is transformed to:

$$\langle\sigma_{+o}\rangle = -\frac{(\omega^{*}_R/2)\langle\sigma_z\rangle}{(i\Delta - \beta_2) + \Gamma\langle\sigma_z\rangle} . \tag{11.13}$$

We simplify the calculations by neglecting, in Eq. (11.9), the terms containing $\operatorname{Re}\Gamma$, since Γ is mainly imaginary. Then, under steady state conditions, Eq. (11.9) is transformed into:

$$-\beta_1\left[\langle\sigma_z\rangle + 1\right] - \omega_R\langle\sigma_{+0}(t)\rangle - \omega^{*}_R\langle\sigma_{-0}\rangle = 0 . \tag{11.14}$$

By substituting Eq. (11.13) into Eq. (11.14), we get:

$$\beta_1\left[\langle\sigma_z\rangle + 1\right] = -\frac{|\omega_R|^2}{2}\langle\sigma_z\rangle\left[\frac{1}{i(\Delta + \operatorname{Im}\Gamma)\sigma_z - \beta_2} + C.C.\right]. \tag{11.15}$$

After rearranging terms in Eq. (11.15), we obtain an equation that is cubic in $\langle\sigma_z\rangle$:

$$\beta_1\left[\langle\sigma_z\rangle + 1\right]\left[\beta_2^{\,2} + \left(\langle\sigma_z\rangle\operatorname{Im}\Gamma + \Delta\right)^2\right] = -\beta_2|\omega_R|^2\langle\sigma_z\rangle . \tag{11.16}$$

Eq. (11.16) is a cubic equation in the population inversion. In Ben-Aryeh, Bowden, and England (1986), this equation was solved numerically. The corresponding Figures show the value of $\langle\sigma_z\rangle$ as a function of $|\omega_R|^2$, for different parameters: β_1, β_2, $\operatorname{Im}\Gamma$, and Δ. We generalized the cubic equation to include the effect of $\operatorname{Re}\Gamma$ in some Figures (in case the system deviates from the above approximation). The Figures show two stable solutions, corresponding to high and low values of $\langle \sigma_z \rangle$ as a function of $|\omega_R|^2$ (described in the Figures by solid lines) and unstable solution (described in the Figures by dashed line). As the electric field increases, the system switches from the lower curve to the upper stable branch at a certain high critical point. Conversely, as the electric field decreases, it switches from the upper curve to the lower branch at a lower critical value. These features describe the bistability properties of the system, arising from a feedback mechanism produced by local field effects. The properties of the present IOB may be compared to conventional optical bistability effects (Lugiato 1984).

Eqs. (11.9) and (11.10) in the present work are equivalent to the semi-classical Eqs. (11.25) and (11.26) obtained in (Bowden and Dowling 1993), though expressed with different notation and under the present approximation that $-\Gamma$ is imaginary, as defined in Bowden and Dowling (1993) as ε. Comparing our value of $-\operatorname{Im}\Gamma$, by substituting $\beta = (4|P^2|k^3\lambda^3)/3\hbar$ in our approximate equation (11.16), we obtain:

$$-\operatorname{Im}\Gamma = \frac{3\beta n\lambda^3}{8\pi^2} = \frac{3n}{8\pi^2}\frac{4|P^2|k^3\lambda^3}{3\hbar} = \frac{4\pi}{\hbar}|P^2|n \,. \tag{11.17}$$

We find that our value of $\varepsilon = (4\pi/3\hbar)|P^2|n$ is equal to the semiclassical value multiplied by a factor of 3 (this discrepancy may result from different approximations used in the two approaches or due to a small difference in the physical system considered).

In Benedict et al. (1991) and Malikov and Malyshev (2017), IOB in thin layers of resonant emitters was treated by using similar methods to those presented in Ben-Aryeh, Bowden, and England (1986) and Bowden and Dowling (1993). In particular, the treatment of the local field effect within the Maxwell equations is similar. However, these works (Benedict et al. 1991; Malikov and Malyshev 2017) add new information into various aspects of IOB e.g., deviations from resonance, the role of inhomogeneous broadening in addition to homogeneous broadening, switching times in IOB, and the effect of the real part of the local field parameter in addition to its imaginary value.

11.3. Long sample with retardation and propagation effects

For coherent radiation transmitted through a long sample with dimensions much larger relative to the wavelength, we use Eqs. (11.5) and (11.6) to study the effect of spatially varying parameters on the behavior of the system. The *z*-axis is defined as the direction of propagation of the externally applied field, and the *x*-axis is defined as the direction of polarization. The coefficients $B(i,j)$ are identical to those obtained from classical dipole-dipole interactions (Milonni and Knight 1974). We simplify their expression by assuming that the dipole moments of the two atoms in $B(i,j)$ are given as:

$$\vec{p}_i = \vec{p}_j = P\hat{x}\,. \tag{11.18}$$

To calculate the cooperative interactions of an atom i with all other atoms, we separate the interaction into two parts. First, we consider a spherical region around the atom i with a radius r_0 on the order of a wavelength. Within this sphere, we can ignore retardation effects and use the mean field approximation. Second, we include the interaction with all other atoms located outside this sphere.

In the second part, for calculating the interaction with the distant dipoles located outside this sphere, we should consider retardation and propagation effects. For this purpose, it is convenient to describe the expectation values of the atomic operators as continuous variables. We define $\langle\sigma_{+0}(r,t)\rangle$ and $\langle\sigma_z(r,t)\rangle$, as the complex dipole and the inversion of population per unit volume at point $\vec{r}$, respectively. Including both interactions, Eqs. (5-6) can be written in the new variables as:

$$\frac{d\langle\sigma_z(r,t)\rangle}{dt} = -\beta_1\left[\langle\sigma_z\rangle + n\right] - \left\{\left[2\Theta^{*}(r,t) + \omega_R\right]\langle\sigma_{+0}(r,t)\rangle + C.C.\right\}, \tag{11.19}$$

$$\frac{d\langle\sigma_{+o}(r,t)\rangle}{dt} = (i\Delta - \beta_2)\langle\sigma_{+o}(r,t)\rangle + \left[\Theta(r,t) + \frac{\omega^{*}_R}{2}\right]\langle\sigma_z\rangle\,, \tag{11.20}$$

where $\Theta(r,t)$ is a continuous variable corresponding to the definition of the reaction field $\Theta(r,t)$ given in Eq. (11.7), and it is separated into two components: $\Theta = \Theta_1 + \Theta_2$ corresponding to the contributions within the sphere and outside the sphere.

Using the mean field approximation from Eqs. (11.9) and (11.10), we express the first part of the cooperative interaction as:

$$\Theta_1(r,t) = \langle \sigma_{+o}(r,t) \rangle \Gamma_1 / n \quad . \tag{11.21}$$

After a lengthy calculation given in Appendix B of Ben-Aryeh, Bowden, and England (1986), the value of Γ_1 was given as a function of $R = kr_0$. By neglecting small oscillations, we get:

$$\mathrm{Re}\Gamma_1 = \frac{3\beta n\lambda^3}{8\pi^2} R \quad ; \quad R = \frac{2\pi}{\lambda} r_0 \quad . \tag{11.22}$$

where $\mathrm{Im}\Gamma_1$ is a negative oscillating function of r_0/λ with an average value:

$$\mathrm{Im}\Gamma = -\frac{7}{6}\frac{3\beta n\lambda^3}{8\pi^2} \quad . \tag{11.23}$$

After further calculations for Θ_2, including retardation and propagation effects, we find that Θ_2 is given in the form of the ordinary Maxwell equation in the slowly varying envelope approximation as:

$$\frac{1}{c}\frac{\partial \Theta_2(z,t)}{\partial t} + \frac{\partial \Theta_2(z,t)}{\partial z} = \frac{3\pi\beta}{k^3}\langle \sigma_{+o}(z,t) \rangle \quad . \tag{11.24}$$

By introducing in Eqs. (11.19) and (11.21) the definitions (Cohen, Ben-Aryeh and Mann 1994):

$$\omega_R' = \omega_R + 2\Theta_2^{\Box}(\vec{r});$$

$$\Gamma' = -\theta_1^{\mathrm{Im}}(\vec{r}) / \langle \sigma_{+o}(\vec{r}) \rangle = -i(3/8\pi^2)\frac{7}{6}\lambda^3\beta. \tag{11.25}$$

The equations assume the same form as Eqs. (11.9-10) for the thin sample, with an imaginary value for Γ. It should be noted that the parameters ω_R and Γ used for the thin sample are replaced here by ω'_R and Γ', respectively. In other words, in the long sample, the effect of Θ_2 is to modify the effective value of ω_R. This leads to the important result that the two phases of high and low transitivity may coexist spatially in the material.

11.4. Experiments verifying IOB due to local dipole-dipole interaction.

Mirrorless with intrinsic bistability in dense media was realized in a set of experiments performed by Hehlen and colleagues (Hehlen et al. 1996; Hehlen et al. 1999; Hehlen et al. 1994; Kuditcher et al. 2000; Lüthi et al. 1998). The two-level density matrix model was extended to incorporate new contributions expected from strong dipolar interactions among *excited states.* Pertinent to a variety of systems. Intrinsic bistability was predicted to depend on a combination of ground- and excited-state coupling, and this phenomenon was observed in cooperative pair luminescence of dimeric compounds. The excitation energy is partitioned between the excited atom and a ground state collision partner at a rate that is bilinear in their respective intensities, and therefore *non-linear* with respect to occupation numbers. The reverse process, called cooperative up-conversion, can be efficient and is the process responsible for the observed IOB. As shown in Hehlen et al. (1994), Figures 1-2, the model incorporates the cooperative dynamics of two three-level atoms. The cooperative dynamics involving two three-level atoms give rise to: (a) cross relaxation and (b) pair up-conversion. The dynamic of interest can be treated with a two-level pair model of cooperative processes. This model incorporates ground-state dipolar interactions, which introduce local field corrections to the optical polarization. Excited-state interactions are also included. Of the three states, one participates as an intermediate state. Its role as a virtual state can be preserved within the framework of the two-level model.

After developing the above model, the dynamic equations for the off-diagonal coherence σ_{21} and the inversion $\sigma_3 = \sigma_{22} - \sigma_{11}$ are obtained. The decay of inversion is given by $-\alpha(\sigma_3 - \sigma_{30})^2 - \gamma(\sigma_3 - \sigma_{30})$, where in the absence of an external field, and under the assumption of negligible thermal occupation, $\sigma_{30}=-1$. The equations of motion used in Hehlen et al. (1994, Eq. 1) are:

$$\frac{d\sigma_{21}}{dt}=-i\left(\Delta+\varepsilon\sigma_3\right)\sigma_{21}-\frac{1}{2}i\Omega\sigma_3-\Gamma\sigma_3 \ ;$$
$$\frac{d\sigma_3}{dt}=-i\Omega\left(\sigma_{21}-\sigma^{*}_{21}\right)-\alpha\left(\sigma_3-\sigma_{30}\right)^2-\gamma\left(\sigma_3-\sigma_{30}\right) \tag{11.26}$$

Here $\varepsilon=n\mu^2/s\hbar\varepsilon_0$ is the local field correction term, which has units of frequency, *n* is the number density, μ is the dipole transition moment, *s* is a factor equal to 3 in cubic solids, ε_0 is the permittivity of vacuum, and Γ is the dephasing rate. The detuning of light from the resonance frequency is denoted by $\Delta=\omega_0-\omega$, and $\Omega=\mu E/\hbar$ is the Rabi frequency. σ_{21} is the off-diagonal matrix element. We find that these equations are equivalent to those given for the thin sample in Ben-Aryeh, Bowden, and England (1986) (with a real local field parameter), and also to the equations presented in Bowden and Dowling (1993). It should be noted that different units are used in different works. Additionally, we notice that Eq. (11.26) includes a quadratic relaxation term. Experimental observations of bistable luminescence from Yb^{3+} pairs in various crystals confirm these theoretical predictions and have interesting implications for local field effects in solids and dense, ultracold gases (Hehlen et al. 1996; Hehlen et al. 1999; Hehlen et al. 1994; Lüthi et al. 1998). The optical bistability is intrinsic and not the result of external feedback. A two-level density matrix model was developed, which includes ground- and excited-state interactions and shows that the intrinsic bistability is due to a local field different from the external field, and which is amplified by the nonlinear cooperative up-conversion. Both theoretical and experimental evidence show that increasing the coupling within Yb^{3+} dimers enhances the observability of IOB. The bi-stable region becomes narrower and shifts to lower excitation energies with increasing temperatures and concentration. Intrinsic bistability and hysteresis have been observed in the photoluminescence of bulk glass doped with Yb^{3+} and Tm^{3+} impurities at room temperature (Kuditcher et al. 2000). It was shown that nonlinear optical material with width smaller than a wavelength can display optical bistability when the phase of the complex third-order susceptibility $\chi^{(3)}$ is nearly opposite to that of the linear dielectric constant ε (Yoon et al.

2000). The effect of a linear host medium on bistability was also studied and it was found that near dipole-dipole interaction are enhanced by the presence of the host medium (Crenshaw and Bowden 1996).

References

Ben-Aryeh, Y., C. M. Bowden, and J. C. England. 1986. "Intrinsic optical bistability in a collection of spatially distributed two-level atoms." *Physical Review A* 34: 3917. https://doi.org/10.1103/PhysRevA.34.3917

Ben-Aryeh, Y., C. M. Bowden, and J. C. England. 1987. "Longitudinal special first-order phase transition in a system of coherently driven two-level atoms." *Optics Communications* 61: 147. https://doi.org/10.1016/0030-4018(87)90237-9

Benedict, M. G., V. A. Malyshev, E. D. Trifonov, and A. I. Zaitsev. 1991. "Reflection and transmission of ultrashort light pulses through a thin resonant medium: Local field effects." *Physical Review A* 43: 3845. https://doi.org/10.1103/PhysRevA.43.3845

Bowden, C. M., and J. P. Dowling. 1993. "Near dipole-dipole effects in dense media: Generalized Maxwell-Bloch equations." *Physical Review A* 47: 1247. https://doi.org/10.1103/PhysRevA.47.1247

Bowden, C. M., and C. C. Sung. 1979. "First- and second-order phase transitions in the Dicke model: Relation to optical bistability." *Physical Review A* 19: 2392. https://doi.org/10.1103/PhysRevA.19.2392

Crenshaw, M. E., and C. M. Bowden. 1996. "Local-field effects in a dense collection of two-level atoms embedded in a dielectric medium: Intrinsic optical bistability enhancement and local cooperative effects." *Physical Review A* 53: 1139. https://doi.org/10.1103/PhysRevA.53.1139

Hehlen, M. P., H. U. Güdel, O. Shu, and S. C. Rand. 1996. "Cooperative optical bistability in the dimer system $Cs_3Y_2Br_9$:10% Yb^{3+}." *Journal of Chemical Physics* 104: 1232–1244. https://doi.org/10.1063/1.471532

Hehlen, M. P., A. Kuditcher, S. C. Rand, and S. R. Lüthi. 1999. "Site-selective, intrinsically bistable luminescence of Yb^{3+} ion pairs in $CsCdBr_3$." *Physical Review Letters* 82: 3050. https://doi.org/10.1103/PhysRevLett.82.3050

Hehlen, M. P., H. U. Güdel, O. Shou, J. Rai, S. Rai, and S. C. Rand. 1994. "Cooperative bistability in dense, excited atomic systems." *Physical Review Letters* 73: 1103. https://doi.org/10.1103/PhysRevLett.73.1103

Hopf, F. A., C. M. Bowden, and W. H. Louisell. 1984. "Mirrorless optical bistability with the use of the local field correction." *Physical Review A* 29: 2591. https://doi.org/10.1103/PhysRevA.29.2591

Kuditcher, A., M. P. Hehlen, C. M. Florea, K. W. Winick, and S. C. Rand. 2000. "Intrinsic bistability of luminescence and stimulated emission in Yb- and Tm-doped glass." *Physical Review Letters* 84: 1898. https://doi.org/10.1103/PhysRevLett.84.1898

Lugiato, L. A. 1984. "II. Theory of optical bistability." *Progress in Optics* 21: 69–216. https://doi.org/10.1016/S0079-6638(08)70122-7

Lüthi, S. R., M. P. Hehlen, T. Riedener, and H. U. Güdel. 1998. "Excited-state dynamics and optical bistability in the dimer system $Cs_3Lu_2Br_9$: Yb^{3+}." *Journal of Luminescence* 76–77: 447. https://doi.org/10.1016/S0022-2313(97)00233-0

Malikov, R. F., and V. A. Malyshev. 2017. "Optical bistability and hysteresis of a thin layer of resonant emitters: Interplay of inhomogeneous broadening of the absorption line and the Lorentz local field." *Optics and Spectroscopy* 122: 955–963. https://doi.org/10.1134/S0030400X17060121

Milonni, P. W., and P. L. Knight. 1974. "Retardation in the resonant interaction of two identical atoms." *Physical Review A* 10: 1096. https://doi.org/10.1103/PhysRevA.10.1096

Yoon, Y.-K., R. S. Bennink, R. W. Boyd, and J. E. Sipe. 2000. "Intrinsic optical bistability in a thin layer of nonlinear optical material by means of local field effects." *Optics Communications* 179: 577–580. https://doi.org/10.1016/S0030-4018(99)00739-7

CHAPTER 12

SEPARABILITY AND ENTANGLEMENT OF 2- AND 3,4-LEVEL QUANTUM SYSTEMS

12.1. Introduction

The problems of separability and entanglement become relatively simple for bipartite systems and much more complicated for systems with higher-level dimensions. Therefore, we first discuss these problems for bipartite systems. We will later analyze the special problem of full-separability and bi-separability for 3 (and more) entangled quantum bits (qubits) mixed with white noise.

12.2. Entanglement and separability in a bipartite system

In the simple case of two-qubit density states, it is possible to use the partial transpose (PT) of the density matrix, according to the Peres-Horodecki (P-H) criterion (Horodecki, Horodecki, and Horodecki 1996; Peres 1996). If the PT of the two-qubit density matrix leads to negative eigenvalues of the PT matrix $\rho_{AB}(PT)$, then the density matrix is entangled; otherwise, it is separable. A bipartite system is separable if and only if the density matrix of this system can be written in the form:

$$\rho_{AB} = \sum_j p_j \rho_A^{(j)} \rho_B^{(j)} \quad . \tag{12.1}$$

Here $p_j > 0$ and $\Sigma_j p_j = 1$. The density matrix ρ_{AB} is defined on the Hilbert space $H_A \otimes H_B$, where A and B are the two parts of the bipartite system. The interpretation of such a definition of separability is that the two systems are independent. The summation over j may include a large number of product density matrices. The density matrix for any bipartite system can be given as:

$$\rho_{AB} = \sum_{i,k,j,l} \rho_{ik,jl} \left(|i\rangle\langle j|\right)_A \left(|k\rangle\langle l|\right)_B . \qquad (12.2)$$

Here, $|i\rangle$ and $\langle j|$ are summed over a complete set of states of system A, while $|k\rangle$ and $\langle l|$ are summed over a complete set of states of system B. The coefficients ρ_{AB} are complex numbers. The PT of the density matrix ρ_{AB} is given as:

$$\rho_{AB}(PT) = \sum_{i,l,j,k} \rho_{ik,lj} \left(|i\rangle\langle j|\right)_A \left(|k\rangle\langle l|\right)_B . \qquad (12.3)$$

Horodecki, Horodecki, and Horodecki (1996) showed that the positivity of the PT state is necessary and sufficient for the separability of 2×2 and 2×3 systems.

For a 2-qubit system denoted by A and B, we use the Hilbert-Schmidt (HS) representation of the density matrix (Ben-Aryeh and Mann 2015; Bengtsson and Życzkowski 2006):

$$\rho_{AB} = \frac{1}{4}\left\{ \begin{matrix} (I)_A \otimes (I)_B + (\vec{r}\cdot\vec{\sigma})_A \otimes (I)_B + (I)_A \otimes (\vec{s}\cdot\vec{\sigma})_B \\ + \sum_{m,n=1}^{3} t_{mn} (\sigma_m)_A \otimes (\vec{\sigma}_n)_B \end{matrix} \right\} . \qquad (12.4)$$

Here, I represents the unit 2×2 matrix, the three Pauli matrices are denoted by $\vec{\sigma}$, while $\vec{r}$ and $\vec{s}$ represent 3-dimensional parameter vectors. The normalized 2-qubit density matrix is described by 15 real HS parameters: 6 for $\vec{r}$ and $\vec{s}$, and 9 for t_{mn}. Assuming the conditions $\vec{r} = \vec{s} = 0$, we get the density matrix:

$$4\rho_{AB} = \left\{ (I)_A \otimes (I)_{B+} + \sum_{m,n=1}^{3} t_{mn} (\sigma_m)_A \otimes (\sigma_m)_B \right\} . \qquad (12.5)$$

The use of PT transformation takes a simple form in the HS representation. Suppose that the density matrix is of the form of Eq. (12.1); then, performing a PT with respect to system A involves replacing ρ by:

$$\rho_{AB}\left(PT;A\right)=\sum_{j}p_{j}\left(\rho_{A}{}^{(j)}\right)^{tr}\rho_{B}{}^{(j)}, \tag{12.6}$$

where the superscript Tr denotes the partial transpose of the density matrix. If the density matrix is expressed as in Eq. (12.4) or Eq. (12.5), the PT operation affects only σ_y to $-\sigma_y$. For a two-level system, if $\rho_{AB}(PT;A)$ has a negative eigenvalue, then ρ_{AB} is entangled; otherwise, it is separable. We demonstrate the PT transformation by applying it to the simple density matrix of the Bohm singlet state, given by:

$$\rho_{AB}=\frac{1}{4}\begin{Bmatrix}(I)_A\otimes(I)_B-(\sigma_x)_A\otimes(\sigma_x)_B \\ -(\sigma_y)_A\otimes(\sigma_y)_B-(\sigma_z)_A\otimes(\sigma_z)_B\end{Bmatrix}. \tag{12.7}$$

Using PT transformation with respect to A, we get:

$$\rho_{AB}\left(PT;A\right)=\frac{1}{4}\begin{Bmatrix}(I)_A\otimes(I)_B-(\sigma_x)_A\otimes(\sigma_x)_B \\ +(\sigma_y)_A\otimes(\sigma_y)_B-(\sigma_z)_A\otimes(\sigma_z)_B\end{Bmatrix}. \tag{12.8}$$

The explicit density matrix of Eq. (12.7) is given by:

$$\rho_{AB}=\begin{pmatrix}0&0&0&0\\0&\frac{1}{2}&-\frac{1}{2}&0\\0&-\frac{1}{2}&\frac{1}{2}&0\\0&0&0&0\end{pmatrix}. \tag{12.9}$$

with eigenvalues: $\lambda_1=\lambda_2=\lambda_3=0;\quad \lambda_4=1$.

The explicit density matrix of Eq. (12.8) is given by:

$$\rho_{AB} = \begin{pmatrix} 0 & 0 & 0 & -\frac{1}{2} \\ 0 & \frac{1}{2} & 0 & 0 \\ 0 & 0 & \frac{1}{2} & 0 \\ -\frac{1}{2} & 0 & 0 & 0 \end{pmatrix} \tag{12.10}$$

with eigenvalues $\lambda_1 = \lambda_2 = 1/2,\ \lambda_3 = 1/4, \text{ and } \lambda_4 = -1/4$, such that the negative value of λ_4 confirms that the singlet Bohm state is indeed entangled.

The von Neumann entropy is defined as $S = -Tr(\rho \ln \rho)$ where tr denotes the trace and $\ln$ denotes the natural matrix logarithm. On the basis of the eigenvectors of the density matrix $\rho = \sum_j \lambda_j |j\rangle\langle j|$ the von Neumann entropy is given by:

$$S = -\sum_j \lambda_j \ln \lambda_j \tag{12.11}$$

For the Bohm singlet state, the entropy equals zero. This result holds for any *pure* state ($j = 1$), indicating that in a pure state, we retain all the quantum information. For *mixed* states ($j > 1$), the quantum entropy is larger than zero, indicating that part of the quantum information is lost.

For a bipartite system, we can use the *Schmidt decomposition* (Pathak 2013): Let H_1 and H_2 be Hilbert spaces of dimensions n and m, and assume $n \geq m$. For any vector W in the tensor product $H_1 \otimes H_2$, there exist orthonormal sets $\{u_1, \cdots, u_m\} \subset H_1$ and $\{v_1, \cdots, v_m\} \subset H_2$ such that:

$$w = \sum_{i=1}^{m} \alpha_i u_i \otimes v_i \ , \tag{12.12}$$

where the scalar α_i are real and non-negative. The partial trace of w, with respect to either system A or B, is a diagonal matrix whose non-zero elements are $|\alpha_i|^2$ which indicates that the reduced states on either subsystem have the same spectrum. There is no simple Schmidt decomposition for systems with 3 levels or higher. In Ben-Aryeh, Mann, and Sanders (1999), we discussed the relations between two-level system entanglement and information theory, with particular emphasis on connections to the Shannon index of correlation.

In Ben-Aryeh and Mann (2015), we analyzed explicit forms of two-qubit separable states that satisfy the relation in Eq. (12.1). We showed that for density matrices of the form given in Eq. (12.5), the matrix t_{mn} can be transformed by local rotations into the form:

$$4\rho_{AB} = \left\{ (I)_A \otimes (I)_{B+} + \sum_{i=1}^{3} t_i (\tilde{\sigma}_i)_A \otimes (\tilde{\sigma}_i)_B \right\}, \tag{12.13}$$

where $\tilde{\sigma}_i$ are rotated Pauli matrices. The proof for this equation is based on the singular value decomposition (SVD), which allows us to choose matrices $(O)_A$ and $(O)_B$ such that:

$$\sum_{m,n} (O_{mi})_A t_{mn} (O_{nj})_B = \delta_{ij} t_i \,. \tag{12.14}$$

A generalization of this equation leads to Eq. (12.13). In Ben-Aryeh and Mann (2015), we proved that the condition:

$$\sum_{i=1}^{3} |t_i|^2 \le 1 , \tag{12.15}$$

is necessary and sufficient for the separability of the density matrix in the form of Eq. (12.13). Other authors also obtained this result, as described in Ben-Aryeh and Mann (2015).

The diagonalization of the density matrix of Eq. (12.4) can be achieved by proper Lorentz transformations (Ben-Aryeh, Mann, and Sanders 1999; Verstraete, Audenaert, and De Moor 2001; Verstraete, Dehaene, and De

Moor 2001). We consider ρ and ρ_M to be of the same equivalence class if:

$$\rho \to \rho_M = M\rho M^\dagger \quad ; \quad M = M_A \otimes M_B, \tag{12.16}$$

where M_A and M_B are invertible matrices. Such equivalence preserves the positivity and separability of the density matrix. For density matrices of the form given in Eq. (12.5), M_A and M_B are orthogonal rotation matrices, so that normalization is preserved. For density matrices of the form in Eq. (12.4), M_A and M_B correspond to Lorentz transformations (Ben Aryeh, Mann, and Sanders 1999, Pathak 2013), and the transformed density matrix must be renormalized. These local transformations reduce the number of parameters in the density matrix of Eq. (12.5) to 3 and in the density matrix of Eq. (12.4) to 4.

12.3. Lorentz transformations for two-level systems

The general density matrix of Eq. (12.4) can be written in the HS representation as:

$$4\rho = \sum_{\alpha,\beta-0}^{3} R_{\alpha,\beta} \left(\sigma_\alpha\right)_A \otimes \left(\sigma_\beta\right)_B. \tag{12.17}$$

Here, the summation extends from 0 to 3, σ_0 is the 2×2 identity matrix, and σ_1, σ_2, and σ_3 are the Pauli spin operators. Verstraete et al. (Verstraete, Audenaert, and De Moor 2001; Verstraete, Dehaene, and De Moor 2001) studied how the density matrix is transformed under local quantum operations of the form:

$$\rho' = \left(A \times B\right) \rho \left(A \times B\right)^\dagger. \tag{12.18}$$

They showed that the $4\otimes4$ R matrix can be expressed as:

$$R = L_1 \Sigma L_2^{\,T}. \tag{12.19}$$

Here, L_1 and L_2 are proper local Lorentz transformations, and Σ is either the diagonal form $\Sigma = diag(s_0, s_1, s_2, s_3)$, the general case considered by

Verstraete, Audenaert, and De Moor (2001), where each $s_0 \geq s_1 \geq s_2 \geq |s_3|$, where s_3 is positive or negative, or of the form:

$$\Sigma = \begin{pmatrix} a & 0 & 0 & b \\ 0 & d & 0 & 0 \\ 0 & 0 & -d & 0 \\ c & 0 & 0 & a+c-b \end{pmatrix}. \tag{12.20}$$

Here a, b, c, and d, are real parameters satisfying one of the following four relations:

a. For $b = c = a/2$, ρ_{AB} is always entangled.
b. For the other 3 cases, ρ_{AB} is separable and has a simple form given explicitly in (Verstraete, Audenaert, and De Moor 2001).

The necessary and sufficient condition for separability is:

$$s_1 + s_s + |s_3| \leq s_0 . \tag{12.21}$$

In Ben-Aryeh and Mann (2017), we describe specific examples for such Lorentz transformations that bring the density matrix into the *explicit* form of Eq. (12.1).

12.4. Full separability and bi-separability of 3,4-qubit entangled states mixed with white noise, using partial transpose, witnesses, and explicit (full/bi) separability

A pure state is fully separable (FS) if it is of the form $|\psi^{fs}\rangle = |\psi_1\rangle \otimes |\psi_2\rangle \otimes |\psi_3\rangle \otimes \cdots$. For an n-qubit density matrix, an explicit expression for FS can be given as:

$$\rho^{fs}{}_{abc\cdots} = \sum_j p_j \rho_A{}^{(j)} \otimes \rho_B{}^{(j)} \otimes \rho_C{}^{(j)} \otimes \cdots, \quad p_j \geq 0 \; ; \; \sum_j p_j = 1 , \tag{12.22}$$

where $\rho_A{}^{(j)}, \rho_B{}^{(j)}, \rho_C{}^{(j)}$ are the density matrices of qubits A, B, and C, respectively.

Several types of entanglement are possible for systems with more than 2 qubits. For example, in the case of 3 qubits, one can distinguish between *full separability, bi-separability*, and *genuine entanglement.* A pure state is fully separable if it is of the form $|\psi^{fs}\rangle = |\psi_1\rangle \otimes |\psi_2\rangle \otimes |\psi_3\rangle \otimes \cdots$.

A pure state $|\phi^{bs}\rangle$ is called bi-separable if it is separable with respect to some partition, e.g., one such partition may be given by:

$$\phi^{bs} = \phi_{AB} \otimes \phi_C \quad ; \quad \phi_C = \begin{pmatrix} \cos\frac{\theta_C}{2} \\ e^{i\varphi_C} \sin\frac{\theta_C}{2} \end{pmatrix}, \tag{12.23}$$

where ϕ_{AB} is a pure state of qubits A and B, and ϕ_C is a pure state of qubit C.

A density matrix is bi-separable if it can be written as:

$$\rho^{bs} = \sum_k p_k |\phi_k^{bs}\rangle\langle\phi_k^{bs}| . \tag{12.24}$$

Explicit expressions for bi-separability of 3-qubit (and larger) systems have been analyzed in the literature (Bancal et al. 2011; Ben-Aryeh and Mann 2016; Teh and Reid 2014). A state is genuinely entangled if it is not bi-separable.

A major obstacle in implementing quantum computational processes is the noise that affects the quantum state. White noise is the simplest type of such noise. Also, since there are many sources of quantum noise (Nielsen and Chuang 2000), we expect that in systems with many qubits, total noise will average out, and white noise may represent this average. In the present work, we were interested in finding the conditions for full separability and genuine entanglement of 3 and 4-qubit entangled states mixed with white noise. The specific problem is as follows: assuming an entangled density matrix ρ_{Ent} of *an n*-qubit system, which is mixed with white noise, the resulting state ρ_{Ent} becomes:

$$\rho_{WN} = \left(\frac{1-p}{2^n}\right)(I)_A \otimes (I)_B \otimes (I)_C \cdots + p\rho_{Ent} \quad (12.25)$$

Here, the subscript WN denotes a mixture of white noise with probability $1-p$ and the entangled state with a density matrix ρ_{Ent} with probability p, of an n-qubit density matrix.

The full HS decomposition of an n-qubit density matrix is given by:

$$2^n \rho = (I)_A \otimes (I)_B \otimes (I)_C \cdots + \sum_{a,b,c=0}^{3} R_{a,b,c} (\sigma_a)_A \otimes (\sigma_b)_B \otimes (\sigma_c)_C \otimes \cdots, \ \sigma_0 = I \, ; a,b,c \neq 0,0,0, \quad (12.26)$$

where the subscripts $A,B,C,\cdots$ refer to different qubits, $\sigma's$ for $a,b,c \neq 0$ are the Pauli operators, and σ_0 is the unit 2×2 operator. The HS parameter $R_{a,b,c\cdots}$ is given by $R_{a,b,c\cdots} = Tr(\rho\sigma_a\sigma_b\sigma_c\cdots)$, although in practice, many of these parameters vanish. PT with respect to qubit A (PTA) is obtained by inverting the sign of each term $R_{a,b,c\cdots}$ of Eq. (12.26) if and only if : $(\sigma_a)_A = (\sigma_y)_A$, i.e., $PTA(\sigma_y)_A = -(\sigma_y)_A$. In a similar way, we can operate with PTB, PTC, etc.

When applying the PT to an n-qubit system, if a negative eigenvalue is obtained, we conclude that the state is not fully separable (though it may still be bi-separable). However, if no negative eigenvalue appears, no conclusion can be obtained regarding the non-separability of the state. If the PT does not provide information, a common method for detecting entanglement is using entanglement witnesses (EW).

An operator EW^{fs} is an entanglement witness for non-full separability of a density matrix ρ if it satisfies two criteria given by:

$$Tr\left(EW^{fs} \left|\psi^{fs}\right\rangle\left\langle\psi^{fs}\right|\right) \geq 0, \quad (12.27)$$

where $\left|\psi^{fs}\right\rangle$ is any fully separable state and

$$Tr\left(EW^{fs}\rho\right)<0 \qquad (12.28)$$

An EW for full separability is given by:

$$EW^{fs}=\alpha^{fs}\left(I\right)^{n}-G_{fs}, \qquad (12.29)$$

where, for n-qubit states I and G_{fs} are the unit and a certain Hermitian operator, respectively. By substituting Eq. (12.29) into Eq. (12.27), we get:

$$Tr\left[\left(\alpha^{fs}\left(I\right)^{n}-G_{fs}\right)\left|\psi^{fs}\right\rangle\left\langle\psi^{fs}\right|\right]=\alpha^{fs}-\left\langle\psi^{fs}\right|G_{fs}\left|\psi^{fs}\right\rangle>0;$$
$$\alpha^{fs}=\max\left\langle\psi^{fs}\right|G_{fs}\left|\psi^{fs}\right\rangle. \qquad (12.30)$$

Next, we substitute EW^{fs} from Eq. (12.29) into Eq. (12.28) and exchange ρ to ρ_{WN} given by Eq. (12.25) (representing an entangled density matrix ρ_{Ent} mixed with white noise). Then, for the maximum value in Eq. (12.28), we obtain:

$$\max Tr\left[\left(\alpha^{fs}\left(I\right)^{n}-G_{fs}\right)\left\{\left(\frac{1-p_{crit}{}^{fs}}{2^{n}}\right)\left(I\right)^{(n)}+p_{crit}{}^{fs}\rho_{Ent}\right\}\right]<0 \qquad (12.31)$$

The idea behind this transformation is that for each entangled density matrix ρ_{Ent}, with a corresponding Hermitian operator, we get a critical value $p_{crit}{}^{fs}$ such that for $p>p_{crit}{}^{fs}$ the system is not fully separable. This condition is sufficient but not necessary, as one might find another operator G_{fs} that will give a larger value of $p_{crit}{}^{fs}$. It was shown in our previous works (Ben-Aryeh and Mann 2016; Ben-Aryeh and Mann 2022) that if the density matrix given in the HS decomposition of Eq. (12.26) can be transformed into the form of Eq. (12.22), then the density matrix is fully separable. If this method provides an explicit construction for full separability as a function of p_{crit}, *below* which the density matrix is fully separable, and if this value is equal to p_{crit} obtained by the witness, *above* which the density matrix is not fully separable, then the coincidence of these two critical values gives the optimal value of p_{crit}.

The limiting value for p_{crit}^{fs} is obtained by assuming equality in Eq. (12.31). After straightforward transformations of Eq. (12.31) (Ben-Aryeh and Mann 2022), we get:

$$p_{crti}^{fs} = \frac{\alpha^{fs} - \frac{TrG_{fs}}{2^n}}{Tr\left(\rho_{Ent} G_{fs}\right) - \frac{TrG_{fs}}{2^n}} . \quad (12.32)$$

It is a challenging task to find G_{fs} such that, by calculating α^{fs} according to Eq. (12.30) and applying Eq. (12.32), we obtain the optimal value p_{crit}^{fs} that satisfies both conditions: above this value, the state is not fully separable according to Eq. (12.32), and below it, the state is fully separable according to the explicit calculations.

To implement the condition of Eq. (12.27) for 3-qubits, we follow the condition that any fully separable pure state density matrix ρ_{fs}, can be given as:

$$\rho_{fs} = \frac{\left(I + \vec{l} \cdot \vec{\sigma}\right)_A}{2} \otimes \frac{\left(I + \vec{m} \cdot \vec{\sigma}\right)_B}{2} \otimes \frac{\left(I + \vec{n} \cdot \vec{\sigma}\right)_C}{2} . \quad (12.33)$$

Here, I is the 2×2 unit matrix, A, B, and C refer to the 3 qubits, and $\vec{\sigma} = \sigma_x \hat{x} + \sigma_y \hat{y} + \sigma_z \hat{z}$, where $\sigma_x, \sigma_y, \sigma_z$ are the Pauli matrices. The vectors $\vec{l}, \vec{m}, \vec{n}$ are 3-dimensional unit vectors with components:

$$l_x = \sin(\theta)_A \cos(\phi)_A \; ; \; l_y = \sin(\theta)_A \sin(\phi)_A \; ; \; l_z = \cos(\theta)_A . \quad (12.34)$$

where m_x, m_y, m_z (n_x, n_y, n_z) are obtained from l_x, l_y, l_z by replacing the subscript A with B (and A to C). The HS decomposition of G_{fs} is given by:

$$G_{fs} = \sum_{a,b,c=1}^{3} G_{a,b,c} \left(\sigma_a\right)_A \otimes \left(\sigma_b\right)_B \otimes \left(\sigma_c\right)_C \; ; \; (a,b,c) \neq (0,0,0) . \quad (12.35)$$

After straightforward calculations (Ben-Aryeh and Mann 2022), we get:

$$\alpha^{fs} = \max \left\langle \psi^{fs} \left| G_{fs} \right| \psi^{fs} \right\rangle = \max \left[l_a m_b n_c G_{a,b,c} \right] \tag{12.36}$$

where $G_{a,b,c}$ are the coefficients of G_{fs} in the HS decompositions ($x \leftrightarrow 1$; $y \leftrightarrow 2$; $z \leftrightarrow 3$).

12.5. Bell operators and full-separability witnesses for $|GHZ(3)\rangle, |W(3)\rangle$ and $|Cl_4\rangle$ and states

Bell operators and entangled witnesses are of significant interest since:

a) Bell operators enable us to negate the possibility of a local hidden variable (LHV) model (Braunstein, Mann, and Revzen 1992; Cabello 2002; Clauser et al. 1969; Gisin and Bechmann-Pasquinucci 1998; Horodecki and Horodecki 1996; Klyshko 1993; Żukowski and Brukner 2002).
b) Entanglement witnesses enable us to negate the separability of a quantum state (e.g., Barreiro et al. 2013; Bourennane et al. 2004; Gühne and Hyllus 2003; Gühne and Tóth 2009; Gühne et al. 2002; Huber and Sengupta 2014; Hyllus et al. 2005; Kye 2015; Nagata et al. 2018; Sanpera, Bruß, and Lewenstein 2001; Terhal 2000; Terhal 2002).

We aim to construct Bell operators, denoted $\hat{B}$, and full-separability witnesses EW^{fs} for specific pure states by using *selected terms from* the HS decompositions of $Tr(\hat{B}\rho)$ and of $Tr(EW^{fs}\rho)$, respectively, not necessarily with the same coefficients, but with certain connections between the HS decompositions. For a Bell operator expressed in the HS decomposition, we must determine the classical bound β_{Class}, where for each σ we assume the value +1 or -1, and the classical bound β_{cl} is given by the maximal value for any possible summation of such values. For a given quantum state, we calculate the quantum value $Tr(\rho\hat{B}) = \beta_{Quant}$. If it exceeds the classical bound, i.e., $\beta_{Quant} > \beta_{Class}$, then ρ cannot be described by the LHV model.

A. The $|GHZ(3)\rangle$ state:

The optimal choice for the Bell operator corresponding to $|GHZ(3)\rangle$ in the HS decomposition is (Gühne and Hyllus 2003; Hyllus et al. 2005):

$$\hat{B}_{G3} = (\sigma_x)_A \otimes (\sigma_x)_B \otimes (\sigma_x)_C - (\sigma_x)_A \otimes (\sigma_y)_B \otimes (\sigma_y)_C$$
$$-(\sigma_y)_A \otimes (\sigma_x)_B \otimes (\sigma_y)_C - (\sigma_y)_A \otimes (\sigma_y)_B \otimes (\sigma_x)_C . \tag{12.37}$$

The full density matrix of $|GHZ(3)\rangle$ in the HS decomposition is:

$$8\rho_{G3} = (I)_A \otimes (I)_B \otimes (I)_C + \hat{B}_{G3} + (I)_A \otimes (\sigma_z)_B \otimes (\sigma_z)_C$$
$$+(\sigma_z)_A \otimes (I)_B \otimes (\sigma_z) + (\sigma_z)_A \otimes (\sigma_z)_B \otimes (I)_C . \tag{12.38}$$

Note that $\hat{B}_{G3}$ is identical to the Mermin Bell operator (Mermin 1990). The quantum limit is 4 while the classical bound is 2, so that the ratio between the classical and the quantum limits is $1/2$. Hence, $|GHZ(3)\rangle$ mixed with white noise cannot be described by an LHV model for $p > 1/2$.

We define the entanglement witness for $|GHZ(3)\rangle$ by:

$$EW_{G3}^{fs} = \alpha_{G3}^{fs} (I)_A \otimes (I)_B \otimes (I)_C - \hat{G}_{fs}^{G3};$$
$$\hat{G}_{fs}^{G3} = \hat{B}_{G3} + (\sigma_z)_A \otimes (\sigma_z)_B \otimes (I)_C . \tag{12.39}$$

Note that we added the term $(\sigma_z)_A \otimes (\sigma_z)_B \otimes (I)_C$, to $\hat{B}_{G3}$. While this addition improves the effectiveness of EW_{G3}^{fs} as a witness operator, it invalidates its use as a Bell operator for LHV analysis. The chosen witness operator $\hat{G}_{fs}^{G3}$ satisfies the relation:

$$Tr[\hat{G}_{fs}{}^{G3}] = 0 . \tag{12.40}$$

Since it is composed of products of Pauli matrices, each with vanishing trace, Eq. (12.32) simplifies in this case to:

$$p_{crit}^{\ fs} = \frac{\alpha^{fs}}{Tr\left(\rho_{Ent} G_{fs}\right)} . \tag{12.41}$$

The coefficients of $\hat{G}_{fs}^{G3}$ in the HS decomposition are given by:

$$G_{xxx}^{G3} = 1, G_{xyy}^{G3} = G_{yxy}^{G3} = G_{yyx}^{G3} = -1; \ \ G_{zz}^{G3} = 1 . \tag{12.42}$$

Substituting Eq. (12.42), into Eq. (12.36), we obtain:

$$\alpha_{G3}^{fs} = \max\left[l_x m_x n_x - l_x m_y n_y - l_x m_x n_y - l_y m_y n_x + l_z m_z \right] = 1 . \tag{12.43}$$

Here, the trigonometric functions are defined as in Eq. (12.34) and their permutations. The value of α_{G3}^{fs} is obtained by the maximum of the trigonometric function in Eq. (12.43).

Using the density matrix of $|GHZ(3)\rangle$ given by Eq. (12.38), $\hat{G}_{fs}^{G3}$ defined in Eq. (12.39), and applying the simplified relation of Eq. (12.41), we obtain:

$$p_{crit} = \frac{\alpha_{G3}^{\ fs}}{Tr[\hat{G}_{fs}^{\ G3} \ \rho_{G3}]} = \frac{1}{5} = 0.2 \ ; \ Tr[\hat{G}_{fs}^{\ G3} \ \rho_{G3}] = 5 . \tag{12.44}$$

Therefore, for $p > 0.2$, $\rho_{WN,G3}$ is not fully separable.

In previous studies (Ben-Aryeh and Mann 2016; Ben-Aryeh and Mann 2018), we have shown, by explicitly expressing the density matrix in the form of Eq. (12.22), that $p = 0.2$ is a sufficient condition for full separability of the $|GHZ(3)\rangle$ state mixed with white noise. We also found that this condition is necessary so that $p \leq 0.2$ is both a sufficient and necessary condition for full separability (equivalent results have also been reported in Acín et al. 2001; Dür, Vidal, and Cirac 2000; Schack and Caves 2000).

B. The $|W(3)\rangle = (1/\sqrt{3})(|100\rangle + |010\rangle + |001\rangle)$ state, in the computational basis:

A simple Bell operator for $|W(3)\rangle$ is obtained by choosing *a subset* of the 3-qubit terms of the HS decompositions (comprising 19 terms, as described in Ben-Aryeh and Mann 2016), with a slight modification of numerical coefficients:

$$\hat{B}_{W3}=(\sigma_z)_A\otimes(\sigma_x)_B\otimes(\sigma_x)_C+(\sigma_x)_A\otimes(\sigma_z)_B\otimes(\sigma_x)_z$$
$$+(\sigma_x)_A\otimes(\sigma_x)_B\otimes(\sigma_z)-(\sigma_z)_A\otimes(\sigma_z)_B\otimes(\sigma_z)_C. \tag{12.45}$$

Here, the quantum limit is $\beta_{Quant}=\langle W(3)|\hat{B}_{W3}|W(3)\rangle=3$, while the classical bound is $\beta_{Clas}=2$, giving a ratio of $p_{crit}=(\beta_{clas}/\beta_{Quant})=2/3$. Hence, $|W(3)\rangle$ has no LHV model for $p>2/3$. This is not necessarily the optimal value.

For analyzing the full separability for $|W(3)\rangle$ we define G_{fs}^{W3} by:

$$\hat{G}_{fs}^{W3}=(\sigma_z)_A\otimes(\sigma_x)_B\otimes(\sigma_x)_C+(\sigma_x)_A\otimes(\sigma_z)_B\otimes(\sigma_x)_C$$
$$+(\sigma_x)_A\otimes(\sigma_x)_B\otimes(\sigma_z)_C-(\sigma_z)_A\otimes(\sigma_z)_B\otimes(\sigma_z)_C$$
$$+(\sigma_z)_A\otimes(\sigma_y)_B\otimes(\sigma_y)_C+(\sigma_y)_A\otimes(\sigma_z)_B\otimes(\sigma_y)_C$$
$$+(\sigma_y)_A\otimes(\sigma_y)_B\otimes(\sigma_z)_C \tag{12.46}$$

Here, we added 3 product terms to the Bell operator, which appear in the HS decomposition of $|W(3)\rangle$ (Ben-Aryeh and Mann 2016). The coefficients of $\hat{G}_{fs}^{W3}$ are:

$$G_{zxx}{}^{W3}=G_{zxx}{}^{W3}=G_{xxz}{}^{W3}=G_{zyy}{}^{W3}=G_{yzy}=G_{yyz}=1\ ;$$
$$G_{zzz}=-1 \tag{12.47}$$

Using Eq. (12.36), we get:

$$\alpha_{W3}^{fs}=\max\begin{bmatrix} l_z m_x n_x+l_x m_z n_x+l_x m_x n_z+l_z m_y n_y \\ +l_y m_z n_y+l_y m_y n_z-l_z m_z n_z \end{bmatrix}=1 \tag{12.48}$$

Applying Eq. (12.41) to the case of $|W(3)\rangle$ mixed with white noise, we get:

$$p_{crit}^{W3} = \frac{\alpha_{W3}^{fs}}{\langle W(3)|\hat{G}_{fs}^{W3}|W(3)\rangle} = \frac{1}{5} = 0.2 \quad . \tag{12.49}$$

The value 5 follows from the HS decomposition of the state $|W(3)\rangle$ (Ben-Aryeh and Mann 2016), where the expectation value for $-(\sigma_z)_A \otimes (\sigma_z)_A \otimes (\sigma_z)_A$ is 1, and the remaining 6 expectation values of $\hat{G}_{fs}^{W3}$, contribute a total of $6\times(2/3)=4$ (see Eq. (26) in Ben-Aryeh and Mann 2022).

In Ben-Aryeh and Mann (2018), we provide an explicit description of the $|W(3)\rangle$ state mixed with white noise in the form of Eq. (12.22) for $p=0.2$. Since the value $p_{crit}=0.2$ is both sufficient (by explicit calculation) and necessary (by the full separability witness), it is considered optimal. It is interesting to note that by using the PT transformation, it was shown in Ben-Aryeh and Mann (2016) that the critical value for full separability is $p=0.209589$ which is slightly above the optimal value (see also Szalay 2011).

C. The $2|Cl_4\rangle = |0000\rangle + |1100\rangle + |0011\rangle - |1111\rangle$ state, in the computational basis:

In the HS decomposition, this state is written as:

$$16\rho_{Cl4} = (I)_A \otimes (I)_B \otimes (I)_C \otimes (I)_D$$
$$+ \sum_{a,b,c,d=0}^{3} R_{a,b,c,d} (\sigma_a)_A \otimes (\sigma_b)_B \otimes (\sigma_c)_C \otimes (\sigma_d)_D \ ;$$
$$(a,b,c,d) \neq (0,0,0,0), \tag{12.50}$$

where:

$$R_{1130} = R_{1103} = R_{3011} = R_{0311} = R_{3300} = R_{0033}$$
$$= R_{3333} = R_{1212} = R_{1221} = R_{2112} = R_{2121} = 1;$$
$$R_{2230} = R_{2203} = R_{3022} = R_{0322} = -1. \tag{12.51}$$

A judicious choice of the Bell operator consists of the following 8 terms, from the HS decomposition:

$$\begin{aligned} B_{Cl4} &= \left[(\sigma_x)_A \otimes (\sigma_y)_b + (\sigma_y)_A \otimes (\sigma_x)_B\right] \\ &\otimes \left[(\sigma_x)_C \otimes (\sigma_y)_D + (\sigma_y)_C \otimes (\sigma_x)_D\right] \\ &+ \left[(\sigma_x)_A \otimes (\sigma_x)_B - (\sigma_y)_A \otimes (\sigma_y)_B\right] \\ &\otimes \left[(\sigma_z)_C \otimes (I)_D + (I)_C \otimes (\sigma_z)_D\right]. \end{aligned} \tag{12.52}$$

Calculating β_{Clas} for this Bell operator, we note that the first square brackets in the first and third lines cannot be simultaneously non-zero. Hence, the classical bound is 4, while the quantum value is: $Tr\left(\rho \hat{B}_{Cl4}\right) = 8$. Thus, the Ratio between the classical bound and the quantum value is $1/2$. By mixing the cluster state with white noise with probability p, we find that for $p > 1/2$, the state Cl_4, cannot be described by an LHV model.

In Ben-Aryeh and Mann (2022, Appendix A), an explicit construction was made for the full separability of $|Cl_4\rangle$ mixed with white noise, in the form of Eq. (12.22) for $p = 1/9$. On the other hand, it was shown, using the PT analysis and the entanglement witness for non-full separability (Ben-Aryeh and Mann 2022), that for $p > 1/9$, the $|Cl_4\rangle$ mixed state with white noise is not fully separable. Hence, the critical value is $p = 1/9$, above which the state is not fully separable, but below it, the state is fully separable.

12.6. Entanglement witness and explicit construction for bi-separability for 3-qubit states and its use for the $|GHZ(3)\rangle$ state

In Eq. (12.23), we described a special form for a 3-qubit bi-separable state. We now aim to represent ϕ_{AB} as a linear combination of four products of two-qubit states:

$$\phi^{bs} = \phi_{AB} \otimes \phi_C \quad ; \quad \phi_C = \begin{pmatrix} \cos\frac{\theta_C}{2} \\ e^{i\varphi_C} \sin\frac{\theta_C}{2} \end{pmatrix} \quad ; \quad \phi_{AB} = \sum_{i=1}^{4} a_i \phi_{AB}{}^{(i)}, \tag{12.53}$$

where, in the computational basis:

$$\phi_{AB}{}^{(1)} = \begin{pmatrix} 1 \\ 0 \end{pmatrix}_A \otimes \begin{pmatrix} 1 \\ 0 \end{pmatrix}_B = \begin{pmatrix} 1 \\ 0 \\ 0 \\ 0 \end{pmatrix} \quad ; \quad \phi_{AB}{}^{(2)} = \begin{pmatrix} 1 \\ 0 \end{pmatrix}_A \otimes \begin{pmatrix} 0 \\ 1 \end{pmatrix}_B = \begin{pmatrix} 0 \\ 1 \\ 0 \\ 0 \end{pmatrix};$$
$$\phi_{AB}{}^{(3)} = \begin{pmatrix} 0 \\ 1 \end{pmatrix}_A \otimes \begin{pmatrix} 1 \\ 0 \end{pmatrix}_B = \begin{pmatrix} 0 \\ 0 \\ 1 \\ 0 \end{pmatrix} \quad ; \quad \phi_{AB}{}^{(4)} = \begin{pmatrix} 0 \\ 1 \end{pmatrix}_A \otimes \begin{pmatrix} 0 \\ 1 \end{pmatrix}_B = \begin{pmatrix} 0 \\ 0 \\ 0 \\ 1 \end{pmatrix}. \tag{12.54}$$

The parameters a_i are the components of a 4-dimensional vector satisfying the relation $\Sigma_{i=1}^{4} |a_i|^2 = 1$. A general representation of these parameters can be given as:

$$a_1 = \cos\theta_1 \; ; \; a_2 = e^{i\varphi_2} \sin\theta_1 \cos\theta_2 \; ; \; a_3 = e^{i\varphi_3} \sin\theta_1 \sin\theta_2 \cos\theta_3 \; ;$$
$$a_4 = e^{i\varphi_4} \sin\theta_1 \sin\theta_2 \sin\theta_3 . \tag{12.55}$$

Here, we used a special form of bi-separability for the partition AB against C, but other forms of bistability also exist, i.e., AC against B or BC against A. Due to the symmetry of the $|GHZ(3)\rangle$ state (including permutation between the 3 qubits), using another form will give the same results as those developed here. By substituting Eqs. (12.54) and (12.55) in Eq. (12.53), we get:

$$\left|\phi^{bs}\right\rangle = \begin{pmatrix} a_1 \cos\theta_C / 2 \\ e^{i\phi_C} a_1 \sin\theta_C / 2 \\ a_2 \cos\theta_C / 2 \\ e^{i\phi_C} a_2 \sin\theta_C / 2 \\ a_3 \cos\theta_C / 2 \\ e^{i\phi_C} a_3 \sin\theta_C / 2 \\ a_4 \cos\theta_C / 2 \\ e^{i\phi_C} a_4 \sin\theta_C / 2 \end{pmatrix}. \tag{12.56}$$

In analogy to Eqs. (12.27)-(12.30), we define the bi-separability witness as:

$$E_{wit}^{bs} = \alpha^{bs} \left(I\right)^n - G_{bs}. \tag{12.57}$$

Here, for an n-qubit system, I is the identity operator of dimension 2^n, G_{bs} is a Hermitian parameter, and α^{bs} is a certain parameter. The superscripts and subscripts *bs* refer to bi-separable states. If, for any $\left|\phi^{bs}\right\rangle$ the following two equations are obeyed:

$$Tr\left(E_{wit}{}^{bs} \left|\phi^{bs}\right\rangle\left\langle\phi^{bs}\right|\right) \geq 0 \; \left(\Rightarrow \alpha^{bs} = \max Tr\left(\left|\phi^{bs}\right\rangle\left\langle\phi^{bs}\right| G_{bs}\right)\right), \tag{12.58}$$

and

$$Tr\left(E_{wit}{}^{bs} \rho\right) < 0, \tag{12.59}$$

then, ρ is genuinely entangled.

In analogy to Eqs. (12.32) and (12.41), we obtain for *bs* states the limiting critical value p_{crit}^{bs}, above which the system is genuinely entangled:

$$p_{crit}^{bs} = \frac{\alpha^{bs} - \frac{TrG_{bs}}{2^n}}{Tr(\rho_{Ent} G_{bs}) - \frac{TrG_{bs}}{2^n}} . \tag{12.60}$$

Under the condition that G_{bs} is traceless:

$$p_{crit}^{bs} = \frac{\alpha^{bs}}{Tr(\rho_{Ent} G_{bs})} . \tag{12.61}$$

12.7. The use of entanglement witness and explicit construction of the bi-separability of $|GHZ(3)\rangle$ state

By combining Eqs. (12.25) and (12.23), the density matrix with probability p mixed with white noise with probability $1-p$, can be given as:

$$\begin{gathered} 8\rho_{wn}^{G3} = (I)_A \otimes (I)_B \otimes (I)_C + pR \\ R = \left[\sum_{a,b,c=0}^{3} R_{a,b,c}^{G3} (\sigma_a)_A \otimes (\sigma_b)_B \otimes (\sigma_c)_C \right]; \\ (a,b,c) \neq (0,0,0,). \end{gathered} \tag{12.62}$$

Assuming that the terms in the square brackets of Eq. (12.62) can be expressed as a sum of bi-separable density matrices minus $g(I)_A \otimes (I)_B \otimes (I)_C$, where g is a certain coefficient. The total coefficient of $(I)_A \otimes (I)_B \otimes (I)_C$ on the right-hand side of Eq. (12.62) becomes $1-pg$. Since this coefficient should be non-negative, we obtain the condition:

$$p_{crit}^{bs} \leq 1/g , \tag{12.63}$$

Under which the state is bi-separable (i.e., not genuinely entangled).

We note that, due to the symmetry of the $|GHZ(3)\rangle$ state, only one form of bi-separability is needed in our proofs. In more general cases of 3-qubit states, analogous equations can be obtained, but all 3 forms of bi-separability must be considered in the general proofs.

A possible choice for the witness operator E_{wit}^{G3} is given by choosing:

$$G_{bs}^{G3} = 8\rho_{GHZ,3} - (I)_A \otimes (I)_B \otimes (I)_C . \tag{12.64}$$

This is a traceless operator, where $\rho_{GHZ,3}$ is the density matrix of the $|GHZ(3)\rangle$ state. According to Eqs. (12.58), and (12.64), we get:

$$\begin{aligned}\alpha_{bs} &= \max\langle\phi_{bs}|\hat{G}_{bs}|\phi_{bs}\rangle = \max\left[8\langle\phi_{bs}|\rho_{GHZ,3}|\phi_{bs}\rangle - 1\right] \\ &= \max\left[8\left|\langle GHZ,3|\phi_{bs}\rangle^2\right| - 1\right].\end{aligned} \tag{12.65}$$

Using Eqs. (12.55), (12.56) and (12.65), we get:

$$\begin{aligned}\sqrt{2}\left|\langle GHZ,3|\phi_{bs}\rangle\right| &\le \left|a_1 \cos\left(\frac{\theta_C}{2}\right) + a_4 e^{i\varphi_C} \sin\left(\frac{\theta_C}{2}\right)\right| \\ &\le \left|\cos\theta_1 \cos\left(\frac{\theta_C}{2}\right)\right| + \left|\sin\theta_1 \sin\left(\frac{\theta_C}{2}\right)\right| \le 1.\end{aligned} \tag{12.66}$$

Hence,

$$\alpha^{bs} = 4 - 1 = 3 . \tag{12.67}$$

Using Eq. (12.61), we then obtain:

$$p_{crit}^{bs,G3} = \frac{\alpha_{bs}^{G3}}{Tr[\hat{G}_{GHZ,3}\rho_{Ent}]} = \frac{3}{7} . \tag{12.68}$$

This result is equivalent to that obtained by other authors (Gühne and Tóth 2009). In order to show that this result is optimal, we showed in Ben-Aryeh and Mann (2022, Appendix B) that an explicit construction of the bi-separability of $|GHZ(3)\rangle$ state mixed with white noise gives the same result, i.e., that ρ_{wn}^{G3} is bi-separable for $p \leq 3/7$.

References

Acín, A., D. Bruß, M. Lewenstein, and A. Sanpera. 2001. "Classification of Mixed Three-Qubit States." *Physical Review Letters* 87 (4): 040401. https://doi.org/10.1103/PhysRevLett.87.040401.

Bancal, J.-D., N. Gisin, Y.-C. Liang, and S. Pironio. 2011. "Device-Independent Witnesses of Genuine Multiparticle Entanglement." *Physical Review Letters* 106 (25): 250404. https://doi.org/10.1103/PhysRevLett.106.250404.

Barreiro, J. T., J.-D. Bancal, P. Schindler, D. Nigg, M. Hennrich, T. Monz, N. Gisin, and R. Blatt. 2013. "Demonstration of Genuine Multipartite Entanglement with Device-Independent Witnesses." *Nature Physics* 9 (9): 559–562. https://doi.org/10.1038/nphys2705.

Ben-Aryeh, Y., A. Mann, and B. C. Sanders. 1999. "Empirical State Determination of Entangled Two-Level Systems and Its Relation to Information Theory." *Foundations of Physics* 29 (12): 1963–1975. https://doi.org/10.1023/A:1018850602867.

Ben-Aryeh, Y., and A. Mann. 2015. "Explicit Constructions of All Separable Two-Qubits Density Matrices and Related Problems for Three-Qubits Systems." *International Journal of Quantum Information* 13 (8): 1550061. https://doi.org/10.1142/S0219749915500616.

Ben-Aryeh, Y., and A. Mann. 2016. "Bi-Separability of 3-Qubits Density Matrices Using Hilbert-Schmidt Decompositions, Sufficient Conditions, and Explicit Expressions." *International Journal of Quantum Information* 14 (7): 1650044. https://doi.org/10.1142/S0219749916500441.

Ben-Aryeh, Y., and A. Mann. 2016. "Separability and Entanglement of n-Qubit and a Qubit and a Qudit, Using Hilbert-Schmidt Decomposition." *International Journal of Quantum Information* 14 (6): 1650030. https://doi.org/10.1142/S0219749916500301.

Ben-Aryeh, Y., and A. Mann. 2017. "Separability and Entanglement of Two Qubits Density Matrices Using Lorentz Transformations." *International Journal of Quantum Information* 15 (7): 1750044. https://doi.org/10.1142/S0219749917500447.

Ben-Aryeh, Y., and A. Mann. 2018. "Separability of 3-Qubits Density Matrices Related to l_1 and l_2 Norms and to Unfolding of Tensors into Matrices." *Journal of Physics Communications* 2 (12): 125011.

https://doi.org/10.1088/2399-6528/aaf1d8.

Ben-Aryeh, Y., and A. Mann. 2022. "Full-Separability and Bi-Separability of Qubits, Using Bell Operators, Partial Transpose, Witnesses and Explicit (Full/Bi) Separability." *International Journal of Quantum Information* 20 (2): 2250011. https://doi.org/10.1142/S0219749922500113.

Bengtsson, I., and K. Życzkowski. 2006. *Geometry of Quantum States: An Introduction to Quantum Entanglement*. Cambridge: Cambridge University Press. https://doi.org/10.1017/CBO9780511535048.

Bourennane, M., M. Eibl, C. Kurtsiefer, S. Gaertner, H. Weinfurter, O. Gühne, P. Hyllus, D. Bruß, M. Lewenstein, and A. Sanpera. 2004. "Experimental Detection of Multipartite Entanglement Using Witness Operators." *Physical Review Letters* 92 (8): 087902. https://doi.org/10.1103/PhysRevLett.92.087902.

Braunstein, S. L., A. Mann, and M. Revzen. 1992. "Maximal Violation of Bell Inequalities for Mixed States." *Physical Review Letters* 68 (22): 3259–3261. https://doi.org/10.1103/PhysRevLett.68.3259.

Cabello, A. 2002. "Bell's Theorem with and without Inequalities for the Three Qubits Greenberger-Horne-Zeilinger and W States." *Physical Review A* 65 (3): 032108. https://doi.org/10.1103/PhysRevA.65.032108.

Clauser, J. F., M. A. Horne, A. Shimony, and R. A. Holt. 1969. "Proposed Experiment to Test Local Hidden-Variable Theories." *Physical Review Letters* 23 (15): 880–884. https://doi.org/10.1103/PhysRevLett.23.880.

Dür, W., G. Vidal, and J. I. Cirac. 2000. "Three Qubits Can Be Entangled in Two Inequivalent Ways." *Physical Review A* 62 (6): 062314. https://doi.org/10.1103/PhysRevA.62.062314.

Gisin, N., and H. Bechmann-Pasquinucci. 1998. "Bell Inequality, Bell States and Maximally Entangled States for n-Qubits." *Physics Letters A* 246 (1–2): 1–6. https://doi.org/10.1016/S0375-9601(98)00516-7.

Gühne, O., and G. Tóth. 2009. "Entanglement Detection." *Physics Reports* 474 (1–6): 1–75. https://doi.org/10.1016/j.physrep.2009.02.004.

Gühne, O., and P. Hyllus. 2003. "Investigating Three Qubit Entanglement with Local Measurements." *International Journal of Theoretical Physics* 42 (5): 1001–1013. https://doi.org/10.1023/A:1025422606845.

Gühne, O., P. Hyllus, D. Bruß, A. Ekert, M. Lewenstein, C. Macchiavello, and A. Sanpera. 2002. "Detection of Entanglement with Few Local Measurements." *Physical Review A* 66 (6): 062305. https://doi.org/10.1103/PhysRevA.66.062305.

Horodecki, M., P. Horodecki, and R. Horodecki. 1996. "Separability of Mixed States: Necessary and Sufficient Conditions." *Physics Letters A* 223 (1–2): 1–8. https://doi.org/10.1016/S0375-9601(96)00706-2.

Horodecki, R., and P. Horodecki. 1996. "Perfect Correlations in the Einstein-Podolsky-Rosen Experiment, and Bell's Inequalities." *Physics Letters A* 210 (4): 227–230. https://doi.org/10.1016/0375-9601(95)00905-1.

Huber, M., and R. Sengupta. 2014. "Witnessing Genuine Multipartite Entanglement with Positive Maps." *Physical Review Letters* 113 (10): 100501. https://doi.org/10.1103/PhysRevLett.113.100501.

Hyllus, P., O. Gühne, D. Bruß, and M. Lewenstein. 2005. "Relations Between Entanglement Witnesses and Bell Inequalities." *Physical Review A* 72 (1): 012321. https://doi.org/10.1103/PhysRevA.72.012321.

Klyshko, D. N. 1993. "The Bell and GHZ Theorems: A Possible Three-Photon Interference Experiment and the Question of Non-Locality." *Physics Letters A* 172 (6): 399–403. https://doi.org/10.1016/0375-9601(93)90228-R.

Kye, S.-H. 2015. "Three-Qubit Entanglement Witness with the Full Spanning Properties." *Journal of Physics A: Mathematical and Theoretical* 48 (23): 235303. https://doi.org/10.1088/1751-8113/48/23/235303.

Mermin, N. D. 1990. "Extreme Quantum Entanglement in a Superposition of Macroscopically Distinct States." *Physical Review Letters* 65 (15): 1838–1840. https://doi.org/10.1103/PhysRevLett.65.1838.

Nagata, K., T. Nakamura, J. Batle, S. Abdalla, and A. Farouk. 2018. "Better Entanglement Witness for Genuine Multipartite Entanglement." *International Journal of Theoretical Physics* 57 (7): 2116–2120. https://doi.org/10.1007/s10773-018-3735-1.

Nielsen, M. A., and I. L. Chuang. 2000. *Quantum Computation and Quantum Information*. Cambridge: Cambridge University Press.

Pathak, A. 2013. *Elements of Quantum Computation and Quantum Communication*. Boca Raton, FL: CRC Press. https://doi.org/10.1201/b15007.

Peres, A. 1996. "Separability Criterion for Density Matrices." *Physical Review Letters* 77 (8): 1413–1415. https://doi.org/10.1103/PhysRevLett.77.1413.

Sanpera, A., D. Bruß, and M. Lewenstein. 2001. "Schmidt-Number Witnesses and Bound Entanglement." *Physical Review A* 63 (5): 050301(R). https://doi.org/10.1103/PhysRevA.63.050301.

Schack, R., and C. M. Caves. 2000. "Explicit Product Ensembles of Separable Quantum States." *Journal of Modern Optics* 47 (2–3): 387–399. https://doi.org/10.1080/09500340008244050.

Szalay, S. 2011. "Separability Criteria for Mixed Three-Qubit States." *Physical Review A* 83 (6): 062337. https://doi.org/10.1103/PhysRevA.83.062337.

Teh, R. Y., and M. D. Reid. 2014. "Criteria for Genuine *N*-Particle Continuous-Variable Entanglement and Einstein-Podolsky-Rosen Steering." *Physical Review A* 90 (6): 062337. https://doi.org/10.1103/PhysRevA.90.062337.

Terhal, B. M. 2000. "Bell Inequalities and the Separability Criterion." *Physics Letters A* 271 (5–6): 319–326. https://doi.org/10.1016/S0375-9601(00)00301-1.

Terhal, B. M. 2002. "Detecting Quantum Entanglement." *Theoretical Computer Science* 287 (1): 313–335. https://doi.org/10.1016/S0375-9601(00)00401-1.

Verstraete, F., J. Dehaene, and B. De Moor. 2001. "Local Filtering Operations on Two Qubits." *Physical Review A* 64 (1): 010101. https://doi.org/10.1103/PhysRevA.64.010101.

Verstraete, F., K. Audenaert, and B. De Moor. 2001. "Maximally Entangled Mixed States of Two Qubits." *Physical Review A* 64 (1): 012316. https://doi.org/10.1103/PhysRevA.64.012316.

Żukowski, M., and Č. Brukner. 2002. "Bell's Theorem for General N-Qubit States." *Physical Review Letters* 88 (21): 210401. https://doi.org/10.1103/PhysRevLett.88.210401.

CHAPTER 13

PSEUDO-HERMITIAN HAMILTONIANS AND BIORTHOGONAL SCALAR PRODUCTS IN POLARIZATION OPTICS

A basic postulate of quantum mechanics is that observables are represented by Hermitian operators on a Hilbert space, and that their time propagators are governed by unitary operators. However, Bender et al. (Bender and Boettcher 1998; Bender, Berry, and Aikaterini 2002; Bender, Brody, and Jones 2002; Bender, Meisinger, and Wang 2003) have suggested replacing the condition of self-adjointness by the weaker condition of parity-time (*pt*) symmetry. This approach often leads to real eigenvalues of the non-Hermitian Hamiltonian. Throughout this chapter, we denote parity-time as *pt* (lowercase) to distinguish it from partial-transpose, which is denoted as *PT* (uppercase).

By introducing an additional symmetry operator, *C* (the conjugation operator), a positive-definite inner product can be constructed (Bender, Brody, and Jones 2002; Bender, Meisinger, and Wang 2003). The concept of a pseudo-Hermitian property was introduced (Mostafazadeh 2002a; Mostafazadeh and Batal 2004), demonstrating that many of the interesting properties of the *pt*-symmetric Hamiltonians follow from their pseudo-Hermitian nature. Pseudo-Hermiticity is defined by a 'distorted' inner product $\langle\psi|\eta\psi\rangle$ where η is a metric operator. Hamiltonian is called pseudo-Hermitian if (Mostafazadeh and Batal 2004; Mostafazadeh 2007):

$$\eta H \eta^{-1} = H. \tag{13.1}$$

Pseudo-Hermiticity is a general condition on a Hamiltonian that allows for the existence of real eigenvalues, and its relation with *pt* is of significant importance. Various mathematical properties of pseudo-Hermitian Hamiltonians have been thoroughly investigated (Bender and Boettcher 1998; Bender, Berry, and Aikaterini 2002; Bender, Brody, and Jones 2002;

Bender, Meisinger, and Wang 2003; Mostafazadeh 2002a; Mostafazadeh and Batal 2004; Mostafazadeh 2007; Ahmed 2003). A substantial body of literature has been published on this topic (e.g., Znojil 2001; Ben-Aryeh and Barak 2006; Ben-Aryeh 2008), including works focused on optical applications (El-Ganainy et al. 2018; Makris et al. 2011). In the following analysis, we use pseudo-Hermitian Hamiltonians to treat a certain problem taken from the field of quantum optics (Ben-Aryeh, Mann, and Yaakov 2004).

13.1. Rabi oscillations in an atomic two-level system with a pseudo-Hermitian Hamiltonian

We assume an excited two-level atomic system whose state is given by:

$$|\psi\rangle = C_a|\phi_a\rangle + C_b|\phi_b\rangle , \tag{13.2}$$

where a and b denote the upper and lower energy levels, respectively. C_a and C_b are the time-dependent amplitudes for these two levels, under the rotating wave approximation and on resonance as follows (Mostafazadeh 2002b):

$$i\frac{dC_a}{dt} = -i\gamma_a C_a + VC_b \quad ; \quad i\frac{dC_b}{dt} = -i\gamma_b C_b + V^{*} C_a \;\; ; \;\; \hbar = 1 \tag{13.3}$$

Here, V is the interaction matrix element, and γ_a and γ_b are the damping constants. We consider a special case under which the Hamiltonian is pseudo-Hermitian. Accordingly, we impose the special conditions:

$$\gamma_a = \gamma \quad ; \quad \gamma_b = -\gamma , \tag{13.4}$$

In this case, γ_b acts as an amplifying constant. Under these special conditions, the time evolution of the state becomes:

$$i\frac{\partial}{\partial t}\begin{pmatrix} C_a \\ C_b \end{pmatrix} = H\begin{pmatrix} C_a \\ C_b \end{pmatrix} \quad ; \quad H = \begin{pmatrix} -i\gamma & V \\ V^{*} & i\gamma \end{pmatrix} \quad ; \quad \hbar = 1 \; . \tag{13.5}$$

The Hamiltonian in Eq. (13.5) is pseudo-Hermitian, as it satisfies the conditions of Theorem 3 in Mostafazadeh (2002b): 'A traceless 2×2 matrix is pseudo-Hermitian if and only if it has a real determinant.'

The Hamiltonian of Eq. (13.5) can be diagonalized, yielding the eigenvalues:

$$E_{\pm} = \pm\sqrt{|V|^2 - \gamma^2} \,. \tag{13.6}$$

The similarity matrix used for this diagonalization of the matrix H, written with simplified notation, is given by:

$$D = \frac{1}{E}\begin{pmatrix} E - i\gamma & -E - i\gamma \\ V^{\bullet} & V^{\square} \end{pmatrix} \;;\; E = E_{+} \;;\; -E = E_{-} \,. \tag{13.7}$$

A corresponding set of eigenvectors is given by:

$$|\psi_1\rangle = \begin{pmatrix} E - i\gamma \\ V^{\square} \end{pmatrix} \;;\; |\psi_2\rangle = \begin{pmatrix} -E - i\gamma \\ V^{\square} \end{pmatrix}. \tag{13.8}$$

We find that under the condition $|V|^2 > \gamma^2$, the energies remain real although the Hamiltonian is non-Hermitian. We adopt the following metric, as proved in Mostafazadeh (2002b):

'If a pseudo-Hermitian $n\times n$ matrix H admits real eigenvalues $(E_1, E_2, \cdots)$, and D is the diagonalization matrix, then H is pseudo-Hermitian where $\eta_{+} = (DD^{\dagger})^{-1}$. The converse is also true.

This theorem does not assume the uniqueness of η_{+}, but we use it to construct a positive-definite metric:

$$\eta_{+} = (DD^{\dagger})^{-1} = \frac{1}{2}\begin{pmatrix} 1 & \frac{iV\gamma}{|V|^2} \\ -\frac{iV^{\square}\gamma}{|V|^2} & 1 \end{pmatrix}. \tag{13.9}$$

By multiplying $|\psi_1\rangle = \begin{pmatrix} E - i\gamma \\ V^{*} \end{pmatrix}$ and $|\psi_2\rangle = \begin{pmatrix} -E - i\gamma \\ V^{*} \end{pmatrix}$ by η_+, we obtain $|\Gamma\rangle_1$ and $|\Gamma\rangle_2$, respectively.

$$\frac{1}{2}\begin{pmatrix} 1 & \frac{iV\gamma}{|V|^2} \\ -\frac{iV^{*}\gamma}{|V|^2} & 1 \end{pmatrix}\left| \begin{matrix} E - i\gamma \\ V^{*} \end{matrix} \right\rangle = \frac{1}{2}\left| \begin{matrix} E \\ V^{*}\left(1 - \frac{i\gamma E + \gamma^2}{|V|^2}\right) \end{matrix} \right\rangle_{1,PH} = |\Gamma\rangle_1$$

$$\frac{1}{2}\begin{pmatrix} 1 & \frac{iV\gamma}{|V|^2} \\ -\frac{iV^{*}\gamma}{|V|^2} & 1 \end{pmatrix}\left| \begin{matrix} -E - i\gamma \\ V^{*} \end{matrix} \right\rangle = \frac{1}{2}\left| \begin{matrix} -E \\ V^{*}\left(1 + \frac{i\gamma E - \gamma^2}{|V|^2}\right) \end{matrix} \right\rangle_{2,PH} = |\Gamma_2\rangle \qquad (13.10)$$

As a consequence of the metric η_+, we obtain the positive-definite inner product:

$$\langle \psi_m | \eta_+ \psi_n \rangle = \langle \psi_m | \Gamma_n \rangle = \delta_{mn} \quad ; \quad |\eta_+ \psi_n\rangle = |\Gamma_n\rangle . \qquad (13.11)$$

The initial two-level wave function is given by:

$$|\psi(0)\rangle = C_1(0)|\psi_1\rangle + C_2(0)|\psi_3\rangle , \qquad (13.12)$$

with time dependence described as:

$$|\psi(t)\rangle = C_1(0)\exp(-iEt)|\psi_1\rangle + C_2(0)\exp(iEt)|\psi_2\rangle . \qquad (13.13)$$

We find that $|\psi(t)\rangle$ is composed of two oscillating terms with frequencies $\pm\sqrt{|V|^2 - \gamma^2}$, corresponding to $|\psi_1\rangle$ and $|\psi_2\rangle$, respectively. However, $|\psi_1\rangle$ and $|\psi_2\rangle$ are biorthogonal states, as defined in Eq. (13.11). Under the condition $|V|^2 - \gamma^2 < 0$, the solutions to the above equations can be expressed as Hyperbolic functions, eliminating oscillatory behavior. Nonetheles, the emphasis of this analysis remains on the existance of real eigenvalues (Bender and Boettcher 1998; Bender, Berry, and Aikaterini 2002; Bender, Brody, and Jones 2002; Bender, Meisinger, and Wang 2003). We also treated the following pseudo-Hermitian Hamiltonian, which appears to be closely related to a realizable physical system (Ben-Aryeh and Barak 2006).

13.2. Pseudo-Hermitian Hamiltonian of a distorted harmonic oscillator

In Ben-Aryeh and Barak (2006), we assumed the simple Hamiltonian:

$$H = \left(p^2 + x^2\right) + ix^3 \exp\left[-\alpha\left(p^2 + x^2\right)\right]. \tag{13.14}$$

Here, $p^2 + x^2$ represents the well-known Harmonic oscillator Hamiltonian, and α is a real, positive parameter. The Hamiltonian defined in Eq. (13.14) is pseudo-Hermitian, and by calculating its eigenvalues (as listed in Table 1 of Ben-Aryeh and Barak 2006), we find that most of them are single valued, with biorthogonality relations under the metric:

$$\eta = \exp\left(-i\pi x\partial / \partial x\right). \tag{13.15}$$

A Hamiltonian H is pseudo-Hermitian if it obeys the following similarity transformation:

$$H = \eta H \eta^{-1} = \exp\left(-i\pi x\partial / \partial x\right) H \exp\left(i\pi x\partial / \partial x\right). \tag{13.16}$$

Since the Hamiltonian is pseudo-Hermitian, it can be given by:

$$H = \sum_{n=0}^{\infty} a_{2n} x^{2n}, \tag{13.17}$$

where the parity transformation $x \to -x$ does not change the term x^{2n}.

Substituting Eqs. (13.15) and (13.17) into Eq. (3.16), we obtain the pseudo-Hermitian relation:

$$\exp(-i\pi x\partial/\partial x)\left(\sum_{n=0}^{\infty}a_{2n}x^{2n}\right)\exp(i\pi x\partial/\partial x)=$$

$$\sum_{n=0}^{\infty}a_{2n}\left[(\exp-i\pi)x\right]^{2n}=\sum_{n=0}^{\infty}a_{2n}x^{2n}=H. \tag{13.18}$$

The equality on the left-hand side of Eq. (13.18) follows from a general operator identity. The purpose of including the exponential term $\exp[-\alpha(p^2+x^2)]$ (which is Pseudo-Hermitian) in Eq. (13.13) is to prevent divergence at very large values of x. We find that large real eigenvalues increase and tend toward those of a standard harmonic oscillator. This tendency becomes more rapid as the parameter α increases.

The eigenvalues of the Hamiltonian in Eq. (13.14) include several pairs of complex conjugate values, with corresponding biorthogonal relations (Ben-Aryeh and Barak 2006). Such pairs might represent bifurcation of the quantum state. It would be of particular interest to investigate whether the 'distorted' harmonic oscillator of Eq. (13.13) can be realized experimentally, along with its associated physical implications.

13.3. *C*, *pt*, and *Cpt* transformations of the pseudo-Hermitian Hamiltonians

The 'time-reversal' and 'spatial-inversion' transformations are usually defined as $(t,\vec{x})=(-t,\vec{x})$ and $(t,\vec{x})=(t,-\vec{x})$, respectively. The latter parity transformation changes a right-handed spatial coordinate system into a left-handed one, referred to as a 'parity' transformation. Following Bender, Brody, and Jones (2002), the $\hat{p}$ operator is defined for the two-level system as:

$$\hat{p}=\begin{pmatrix}0 & 1\\ 1 & 0\end{pmatrix}, \tag{13.19}$$

so that $\hat{p}=\hat{p}^{-1}$, and $\hat{p}^2=I$. The non-uniqueness of $\hat{p}$ and $\hat{t}$ operators follow from the non-uniqueness of the metric operator. The antilinear operator $\hat{t}$ is described in the general case by the product of the unitary operator U and the complex conjugation operator K_0 (Gottfried 1966):

$$\hat{t} = \hat{U}\hat{K}_0 \quad ; \quad \hat{K}_0^{\,2} = 1 \quad ; \quad \hat{t}^{-1} = K_0\hat{U}^{\dagger} \, . \tag{13.20}$$

Following Bender, Brody, and Jones (2002), we define for our system $\hat{U} = I$, so that $\hat{t} = \hat{K}_0$. In the special case where $\hat{U} \neq I$, we find that the system exhibits an additional symmetry $\hat{K}_0 = \hat{C}$.

The pseudo-Hermitian formalism has been applied to polarization optics (Ben-Aryeh 2005). The following analysis demonstrates how non-unitary transformations and biorthogonal scalar products are implemented in polarization optics.

13.4. Biorthogonal scalar products and non-unitary transformations in polarization optics

It is assumed that a light beam is incident normally on the optical system, propagating along the *z*-axis. The polarization state of the light is described by a Jones column vector (Yariv 1991; Yariv and Yeh 1984):

$$\vec{V} = \begin{pmatrix} V_x \\ V_y \end{pmatrix} . \tag{13.21}$$

Here, V_x and V_y are complex numbers representing the EM field components along the *x*- and *y*-coordinates. The orthogonal basis vectors are:

$$\hat{V}_x = \begin{pmatrix} 1 \\ 0 \end{pmatrix} \quad ; \quad \hat{V}_y = \begin{pmatrix} 0 \\ 1 \end{pmatrix} . \tag{13.22}$$

13.5. Lorentz group algebra SU(1,1) for polarization states

The generators K_1, K_2, and K_3 are given by the 2×2 matrices:

$$K_1 = \frac{i\sigma_z}{2} = \begin{pmatrix} \frac{i}{2} & 0 \\ 0 & \frac{-i}{2} \end{pmatrix} \; ; \; K_2 = \frac{i\sigma_x}{2} = \begin{pmatrix} 0 & \frac{i}{2} \\ \frac{i}{2} & 0 \end{pmatrix} \; ;$$

$$K_3 = \frac{\sigma_y}{2} = \begin{pmatrix} 0 & -\frac{i}{2} \\ \frac{i}{2} & 0 \end{pmatrix}. \tag{13.23}$$

The non-unitary squeezing operators are given by:

$$\exp(2i\gamma K_1) = \begin{pmatrix} \exp(-\gamma) & 0 \\ 0 & \exp(\gamma) \end{pmatrix} \; ;$$

$$\exp(2irK_2) = \begin{pmatrix} \cosh(r) & -\sinh(r) \\ -\sinh(r) & \cosh(r) \end{pmatrix}, \tag{13.24}$$

where γ and r are the squeezing parameters. The rotation matrix is given by:

$$R(\xi) = \exp(2i\xi K_3) = \begin{pmatrix} \cos(\xi) & \sin(\xi) \\ -\sin(\xi) & \cos(\xi) \end{pmatrix}. \tag{13.25}$$

The *partial polarization* is described by:

$$P = \begin{pmatrix} p_1 & 0 \\ 0 & p_2 \end{pmatrix} = \sqrt{p_1 p_2} \begin{pmatrix} (p_1/p_2)^{1/2} & 0 \\ 0 & (p_2/p_1)^{1/2} \end{pmatrix} \; ; \; 0 < p_1, p_2 < 1 \; . \tag{13.26}$$

Since we are interested in the normalized polarization vectors, and not in the isotropic loss, we get:

$$\tilde{P} = \begin{pmatrix} (p_1/p_2)^{1/2} & 0 \\ 0 & (p_2/p_1)^{1/2} \end{pmatrix} = \begin{pmatrix} \exp(-\gamma) & 0 \\ 0 & \exp(\gamma) \end{pmatrix} \; ;$$

$$(p_1/p_2)^{1/2} = \exp(-\gamma), \tag{13.27}$$

which is of the form $\exp(2i\gamma K_1)$.

13.6. Linear transformation for retardation plates

The linear transformation associated with a retardation plate is given by:

$$\begin{pmatrix} V'_x \\ V'_y \end{pmatrix} = U \begin{pmatrix} V_x \\ V_y \end{pmatrix} = \begin{pmatrix} \exp\left(-in_s\omega l / c\right) & 0 \\ 0 & \exp\left(-in_f\omega l / c\right) \end{pmatrix} \begin{pmatrix} V_x \\ V_y \end{pmatrix}, \tag{13.28}$$

where n_s and n_f are the refractive indices of the slow and fast eigenwaves, respectively. The parameter l denotes the thickness of the plate, and ω, is the frequency of the light beam. The phase retardation is defined as half the phase difference of the phase delays:

$$E = \left(n_s - n_f\right)\frac{\omega l}{2c}. \tag{13.29}$$

The mean absolute phase change is given as:

$$\alpha = \left(n_s + n_f\right)\frac{\omega l}{2c}. \tag{13.30}$$

Then, Eq. (13.25) represents a transformation described by the unitary matrix U:

$$U = \exp\left(-i\alpha\right)\exp(iH) \quad ; \quad H = \begin{pmatrix} -E & 0 \\ 0 & E \end{pmatrix}$$

$$U = \exp\left[-i\left(n_s + n_f\right)\frac{\omega l}{2c}\right]\exp\begin{bmatrix} -i\left(n_s - n_f\right)\frac{\omega l}{2c} & 0 \\ 0 & i\left(n_s - n_f\right)\frac{\omega l}{2c} \end{bmatrix}. \tag{13.31}$$

It is important to note that the transformation matrices for partially polarized light become unimodular when neglecting the isotropic losses. The effect of an optical system composed of partial polarizer, rotators, and retardation plates is obtained by multiplying the individual complex unimodular matrices corresponding to the optical matrix elements. The transformation of the Jones column vector $\vec{V}$, is then obtained by the overall unimodular transfer matrix S:

$$S = \begin{pmatrix} s_1 & s_3 \\ s_4 & s_2 \end{pmatrix} \; ; \; s_1 s_2 - s_3 s_4 = 1 \quad . \tag{13.32}$$

13.7. Realization of similarity transformation in polarization optics

Suppose we have a Hamiltonian H with two orthogonal polarization eigenvectors. The general transformation of the Hamiltonian H in polarization optics is given by:

$$\tilde{H} = S^{-1} H S \, , \tag{13.33}$$

where S is defined as in Eq. (13.32). The general transformation by the matrix S may include non-unitary transformations, such as the one described in Eq. (13.27). In such cases, the transformation in Eq. (13.33) leads to a pseudo-Hermitian Hamiltonian, where the polarization eigenvectors are biorthogonal, having a certain metric such as defined in Eq. (13.11), where $\eta = (SS^{\dagger})^{-1}$.

The core idea of this formalism is that we may begin with orthogonal polarization eigenvectors of H , and by applying similarity transformations using invertible matrices S, which can be physically implemented in polarization optics, we can produce a wide variety of pseudo-Hermitian eigenvectors. We note that the pseudo-Hermitian formalism can be realized experimentally in polarization optics (El-Ganainy et al. 2018; Makris et al. 2011; Ben-Aryeh 2005).

References

Ahmed, Z. 2003. "*C*-, *PT*-, and *CPT*- Invariance of Pseudo-Hermitian Hamiltonians." *Journal of Physics A: Mathematical and General* 36: 9711. https://doi.org/10.1088/0305-4470/36/37/309

Ben-Aryeh, Y. 2005. "Nonunitary Squeezing and Biorthogonal Scalar Products in Polarization Optics." *Journal of Optics B: Quantum and Semiclassical Optics* 7: S452–S457.

https://doi.org/10.1088/1464-4266/7/12/004

Ben-Aryeh, Y. 2008. "Quasi-Hermitian Hamiltonians, Minimum-Uncertainty (MU) Angular Momentum States and Interferometers." *Journal of Modern Optics* 55: 261–272. https://doi.org/10.1080/09500340701403152

Ben-Aryeh, Y., A. Mann, and I. Yaakov. 2004. "Rabi Oscillations in a Two-Level Atomic System with a Pseudo-Hermitian Hamiltonian." *Journal of Physics A: Mathematical and General* 37: 12059–12066. https://doi.org/10.1088/0305-4470/37/50/008

Ben-Aryeh, Y., and R. Barak. 2006. "Parity-Time (PT) Invariance and Pseudo-Hermitian Hamiltonians of a Non-Linear Oscillator." *Physics Letters A* 351: 388–390. https://doi.org/10.1016/j.physleta.2005.11.031

Bender, C. M., and S. Boettcher. 1998. "Real Spectra in Non-Hermitian Hamiltonians Having PT Symmetry." *Physical Review Letters* 80: 5243. https://doi.org/10.1103/PhysRevLett.80.5243

Bender, C. M., D. C. Brody, and H. F. Jones. 2002. "Complex Extension of Quantum Mechanics." *Physical Review Letters* 89: 270401. https://doi.org/10.1103/PhysRevLett.89.270401

Bender, C. M., M. V. Berry, and A. Aikaterini. 2002. "Generalized PT Symmetry and Real Spectra." *Journal of Physics A: Mathematical and General* 35: L467. https://doi.org/10.1088/0305-4470/35/31/101

Bender, C. M., P. N. Meisinger, and Q. Wang. 2003. "Calculation of the Hidden Symmetry Operator in PT-Symmetric Quantum Mechanics." *Journal of Physics A: Mathematical and General* 36: 1873–1983. https://doi.org/10.1088/0305-4470/36/7/312

El-Ganainy, R., K. G. Makris, M. Khajavikhan, Z. H. Musslimani, S. Rotter, and D. N. Christodoulides. 2018. "Non-Hermitian Physics and PT Symmetry." *Nature Physics* 14: 11. https://doi.org/10.1038/nphys4323

Gottfried, K. 1966. *Quantum Mechanics: Fundamentals*. New York: Benjamin.

Makris, K. G., R. El-Ganainy, D. N. Christodoulides, and Z. H. Musslimani. 2011. "PT-Symmetric Periodic Optical Potentials." *International Journal of Theoretical Physics* 50: 1019–1041. https://doi.org/10.1007/s10773-010-0625-6

Mostafazadeh, A. 2002a. "Pseudo-Hermiticity versus PT Symmetry: The Necessary Condition for the Reality of the Spectrum of a Non-Hermitian Hamiltonian." *Journal of Mathematical Physics* 43: 205–214. https://doi.org/10.1063/1.1418246

Mostafazadeh, A. 2002b. "Pseudo-Hermiticity for a Class of Non-Diagonalized Hamiltonians." *Journal of Mathematical Physics* 43: 6343. https://doi.org/10.1063/1.1514834

Mostafazadeh, A. 2007. "Time-Dependent Pseudo-Hermitian Hamiltonians Defining a Unitary Quantum System and Uniqueness of the Metric Operator." *Physics Letters B* 650: 208–212. https://doi.org/10.1016/j.physletb.2007.04.064

Mostafazadeh, A., and A. Batal. 2004. "Physical Aspects of Pseudo-Hermitian and PT-Symmetric Quantum Mechanics." *Journal of Physics A: Mathematical and General* 37: 11645–11679. https://doi.org/10.1088/0305-4470/37/48/009

Yariv, A. 1991. *Optical Electronics*. Philadelphia: Saunders.

Yariv, A., and P. Yeh. 1984. *Optical Waves in Crystals*. New York: Wiley.

Znojil, M. 2001. "PT-Symmetric Square Well." *Physics Letters A* 285: 7–10. https://doi.org/10.1016/S0375-9601(01)00301-2

CHAPTER 14

DYNAMICAL AND GEOMETRIC PHASES IN QUANTUM ENGINE SYSTEMS

14.1. Introduction

Quantum thermal engines connected to a hot and cold heat baths have been extensively studied, often based on the previous studies of Otto (Abah et al. 2012; Kosloff and Rezek 2017; Leff 1987; Robnagel et al. 2016), and Carnot (Curzon and Ahlborn 1975; Esposito et al. 2010; Geva and Kosloff 1992; Lin and Chen 2005) cycles. The thermodynamic behavior and efficiency of these engines have been thoroughly analyzed. In the present work, we raise the question of whether interference effects might be observed in such systems.

We distinguish between dynamical and geometric phases (Ben-Aryeh 2004; Berry 1984). If a quantum system is initially in an eigenstate of the Hamiltonian H, and the Hamiltonian changes in time, according to the slowly varying parameters $\vec{R}$, the adiabatic conditions guarantee that this $n'th$ state remains in the instantaneous eigenstate of the Hamiltonian:

$$H(R)\left|n,\vec{R}\right\rangle = E_n\left(\vec{R}\right)\left|n,\vec{R}\right\rangle . \tag{14.1}$$

By assuming that the $\vec{R}$ parameters are slowly varying along a closed path C in parameter space, the geometric phase factor at time T is then given by (Ben-Aryeh 2004; Berry 1984):

$$\gamma_n(C) = i\oint d\vec{R}\cdot\left\langle n,\vec{R}\right|\vec{\nabla}_R\left|n,\vec{R}\right\rangle . \tag{14.2}$$

This phase is acquired in addition to the dynamical phase, given by:

$$\phi_{dyn} = -\int_0^T E_n(t)dt \qquad (14.3)$$

In the present chapter, we find that the analyses on quantum thermal engines are typically performed using expectation values for the dynamical phases, which renders them unsuitable for describing quantum interference effects. Classical geometric phases have been obtained in stochastic chemical kinetics in quantum dots coupled to two reservoirs. These works, as discussed in Ben-Aryeh (2022), refer to *classical* geometric phases, as they are obtained from rate equations for probabilities. Since we are interested in quantum effects, quantum geometric phases will be treated in the last part of this chapter (Ben-Aryeh 2025). We begin by reviewing standard theories of quantum engines.

14.2. Reversible and irreversible harmonic oscillator Otto cycle

We consider a quantum thermal engine whose working medium is a single harmonic oscillator, where the engine operates in 4 steps. In two steps, the time-dependent frequency of the harmonic oscillator changes between ω_1 and ω_2, and vice versa (Abah et al. 2012). In the other two steps, the engine is alternatively coupled to two heat baths at a hot temperature T_H and a cold temperature T_C.

There is a trade-off between maximum power and maximum efficiency, at which power tends to zero. In the following analysis, we treat the limiting case of an Otto cycle operating under reversible conditions. In this case, the efficiency is maximal, but the output power tends to zero.

14.3. Otto cycle, reversible with maximal efficiency

When the parameters are changed infinitely slowly along a path from a certain initial point to a final point in parameter space, the total work W performed on the system is equal to the Helmholtz free energy difference ΔF (Jarzynski 1997). The Helmholtz free energy is defined as $F = U - TS$, where U is the internal energy, T is the absolute temperature, and S is the entropy. The Otto cycle consists of four consecutive steps (see Figure 1 of Abah et al. 2012).

1. In the first step, an isentropic compression occurs. The frequency of the oscillator changes between points A and B, from ω_1 to ω_2, while the temperature T_H remains unchanged. The mean work performed in this step is given by:

$$\langle W_1 \rangle = \langle H \rangle_B - \langle H \rangle_A = \left(\frac{\hbar\omega_2}{2} - \frac{\hbar\omega_1}{2} \right) \coth\left(\frac{\beta_1 \hbar\omega_1}{2} \right);$$

$$\beta_1 = \frac{1}{k_b T_H}, \tag{14.4}$$

where k_B is the Boltzmann constant.

2. In the second step, the temperature of the oscillator changes between points B and C, from T_H to T_C, but the frequency ω_2 is unchanged. The mean heat exchanged is given by:

$$\langle Q_2 \rangle = \langle H \rangle_C - \langle H \rangle_B = \frac{\hbar\omega_2}{2} \left[\coth\left(\frac{\beta_2 \hbar\omega_2}{2} \right) - \coth\left(\frac{\beta_1 \hbar\omega_1}{2} \right) \right];$$

$$\beta_2 = \frac{1}{k_b T_C}.$$

(14.5)

3. In the third step, an isentropic expansion occurs. The frequency of the oscillator is changed, and the average work done in this step is given by:

$$\langle W_3 \rangle = \langle H \rangle_D - \langle H \rangle_C = \left(\frac{\hbar\omega_1}{2} - \frac{\hbar\omega_2}{2} \right) \coth\left(\frac{\beta_2 \hbar\omega_2}{2} \right). \tag{14.6}$$

4. In the fourth step, the temperature of the oscillator changes between the points D and A, from T_H to T_C, but its frequency ω_1 is unchanged. The mean heat exchanged in this step is given by:

$$\langle Q_4 \rangle = \langle H \rangle_A - \langle H \rangle_D = \frac{\hbar\omega_1}{2} \left[\coth\left(\frac{\beta_1 \hbar\omega_1}{2} \right) - \coth\left(\frac{\beta_2 \hbar\omega_2}{2} \right) \right].$$

(14.7)

By dividing $-(W_1 + W_3)$ (representing the work obtained in the Otto circuit) by Q_2, given by Eq. (14.5), we get the efficiency of this Otto circuit as:

$$\eta = -\frac{W_1 + W_3}{Q_2} = 1 - \frac{\omega_1}{\omega_2}. \quad (14.8)$$

14.4. Otto cycle at high temperature

We consider the high-temperature limit, for which $\beta_j \hbar \omega_j \ll 1 \ (j = 1,2)$. Then, we obtain the following approximations from Eqs. (14.4) and (14.6):

$$\langle W_1 \rangle = (\hbar\omega_2 / 2 - \hbar\omega_1 / 2)\frac{2}{\beta_1 \hbar \omega_1};$$
$$\langle W_3 \rangle = (\hbar\omega_1 / 2 - \hbar\omega_2 / 2)\frac{2}{\beta_2 \hbar \omega_2}, \quad (14.9)$$

$$2\frac{W_1 + W_3}{\hbar} = \frac{1}{\beta_1}\left(\frac{\omega_2}{\omega_1} - 1\right) + \frac{1}{\beta_2}\left(\frac{\omega_1}{\omega_2} - 1\right). \quad (14.10)$$

Assuming fixed values for β_1 and β_2, the maximal value of $W_1 + W_3$ is obtained by setting the derivative of Eq. (14.10) with respect to ω_2 or ω_1, equal to zero. Then we get:

$$\frac{\omega_2}{\omega_1} = \sqrt{\frac{\beta_1}{\beta_2}}. \quad (14.11)$$

Substituting Eq. (14.11) in Eq. (14.8), we get the efficiency at maximum power:

$$\eta = 1 - \sqrt{\frac{\beta_2}{\beta_1}} = 1 - \sqrt{\frac{T_C}{T_H}}, \quad (14.12)$$

which corresponds to Curzon-Ahlborn efficiency (Curzon and Ahlborn 1975).

The efficiency of an irreversible Otto cycle at maximum power, with proposed realization in a Paul trap, was analyzed in Abah et al. (2012). In the following analysis, we will use conventional methods for treating quantum engines operating with spin systems. Later, we will examine the possibility of obtaining geometric phases in these quantum engine systems, with implications for interference effects.

14.5. Quantum thermal engine with spin-1/2 system

A quantum Carnot engine is described, in which the working medium is a two-dimensional spin-1/2 system subjected to a time-dependent magnetic field in the symmetric *z*-direction (Geva and Kosloff 1992; Lin and Chen 2005). The dynamic of this engine is obtained in four steps. In two of the steps, the system is alternately coupled to hot and cold heat baths. In the other two steps, the time development is adiabatic and isentropic (with constant entropy).

The magnetic field $\vec{B}$ is along the positive *z*-axis, and the Hamiltonian is given by:

$$\hat{H}(t) = 2\mu_B B_z(t)\hat{s}_z \equiv \tilde{\omega}(t)\hat{s}_z \quad ; \quad \tilde{\omega}(t) = 2\mu_B B_z(t) \,. \tag{14.13}$$

The units are such that $\hbar = 1$, μ_B is the Bohr magneton, $\tilde{\omega}(t)$ is proportional to $B_z(t)$, and $\hat{s}_z$ is the spin operator in the *z*-direction. We refer to $\tilde{\omega}(t)$, rather than $B_z(t)$, as "the field".

The mixed density matrix ρ_0 of this system is given by:

$$\rho_0 = P_1|-1/2\rangle\langle -1/2| + P_2|+1/2\rangle\langle +1/2| \,. \tag{14.14}$$

The values $\{-1/2, +1/2\}$ are the eigenvalues of the spin, and the classical probabilities P_1 and P_2 are given, respectively, by statistical mechanics as:

$$P_1 = \frac{\exp(\hbar\tilde{\omega}\beta/2)}{\exp(\hbar\tilde{\omega}\beta/2)+\exp(-\hbar\tilde{\omega}\beta/2)};$$
$$P_2 = \frac{\exp(-\hbar\tilde{\omega}\beta/2)}{\exp(\hbar\tilde{\omega}\beta/2)+\exp(-\hbar\tilde{\omega}\beta/2)};$$
$$\beta = \frac{1}{k_B T}, \tag{14.15}$$

where $\tilde{\omega}$ was defined in Eq. (14.13). Based on statistical mechanics, the expectation value of $\hat{s}_z$ is given by:

$$\langle s_z \rangle = -\frac{1}{2}\tanh(\tilde{\omega}\beta/2); \quad \beta = \frac{1}{k_b T}, \tag{14.16}$$

and the expectation value of the Hamiltonian is given by:

$$E_0 = \langle H_0 \rangle = \hbar\tilde{\omega}\langle s_z \rangle . \tag{14.17}$$

Here, H_0 is the zero-order Hamiltonian, and E_0 is its expectation value. This quantum machine operates through different steps in which the energy of the spin-1/2 (considered as the working medium) is given by the expectation value of Eq. (14.13). In each step, either the temperature, the "frequency" $\hat{\omega} = 2\mu_B B_z$, or both parameters are varied.

14.6. Reversible and irreversible Carnot cycle with spin-1/2 system

The Carnot engine, including its efficiency, can be analyzed under different limiting cases. In case A, we assume that the parameters of the Carnot cycle are changed *infinitely* slowly. The Carnot quantum engine is then reversible, with maximal efficiency, but the power output tends to zero. In case B, we consider the irreversible Carnot cycle, which gives maximal power but involves energy losses.

14.7. Case A: Reversible Carnot cycle

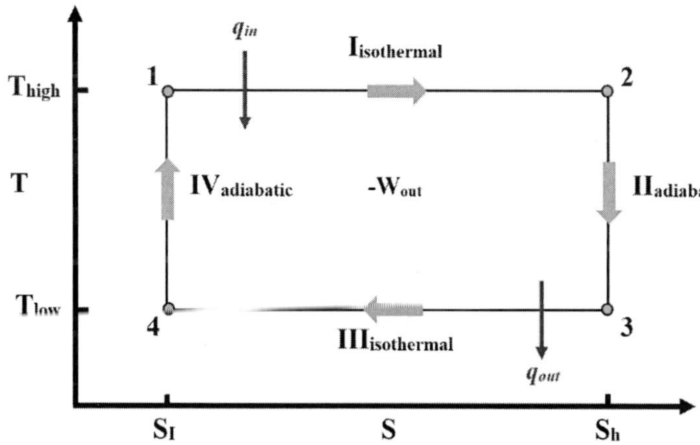

Figure 14.1: Carnot cycle. Reproduced from Ben-Aryeh, Y. 2025. "Quantum Thermal Engine Operating with Spin 1/2 System and Geometric Phases Obtained by Unitary Transformation Matrix of Mixed States." *International Journal of Thermodynamics.* Erken Görünüm-Early Pub Issues: 1-6. Licensed under CC BY).

The Carnot cycle is described using four points, 1, 2, 3, and 4, that denote four different states of the system. The transitions between states 1 and 2, and between states 3 and 4, are described, for the hot isotherm at temperature T_{high}, and for the cold isotherm at temperature T_{low}, respectively. The transitions between states 2 and 3, and that between states 4 and 1 are adiabatic with constant entropy S_h and S_l, respectively.

Carnot's quantum engine operates in 4 steps, schematically illustrated in Figure 14.1. At each of the points 1, 2, 3, and 4, the expectation value of the Hamiltonian in Eq. (14.13) is given as $\hbar\tilde{\omega}\langle s_z\rangle$, where $\tilde{\omega}(t)=2\mu_B B_z(t)$, and $\langle s_z\rangle=-\tanh(\tilde{\omega}\beta/2)/2$ is the expectation value of the spin-1/2 $\hat{s}_z$ operator. During the transitions between consecutive points, the parameter $\tilde{\omega}$ (proportional to the magnetic field) or $\beta=1/k_BT$ (or both) may change.

The transition between points 1 and 2 describes the hot isotherm (denoted in Fig. 1 as $I_{isothermal}$), where the system is coupled to a hot reservoir at temperature T_{high}. In this step, heat q_{in} is entering the system. The transition between points 3 and 4 describes the cold isotherm (denoted in Fig. 1 as $III_{isothermal}$), where the system is coupled to a cold reservoir at temperature T_{low}. In this step, heat q_{out} exits the system. The transition between points 2 and 3 is adiabatic and isentropic (denoted in Fig. 1 as $II_{adiabatic}$ with constant entropy S_h). The transition between 4 and 1 is adiabatic and isentropic (denoted in Fig. 1 as $IV_{adiabatic}$), with constant entropy S_l. In these steps (denoted as $II_{adiabatic}$ and $IV_{adiabatic}$), no heat is transferred, as the system is isolated from the reservoirs. However, work is done by decreasing or increasing the temperature by the difference $\pm(T_H - T_C)$, respectively.

For an ideal Carnot cycle (assuming no losses), the work done by such a thermal machine is given by $q_{in} + q_{out}$, where q_{out} is negative. The Carnot efficiency is then given in the limiting case by $\eta = (q_{in} + q_{out}) / q_{in} = (T_{high} - T_{low}) / T_{High}$. Fig. 1 describes the physical nature of the present Carnot cycle, but it does not provide quantitative experimental values for each step. Explicit evaluations of the experimental values of the present quantum Carnot engine are given below.

The heat q_{in} absorbed by the present spin system is positive and given by:

$$
\begin{aligned}
&q_{in} = \langle H_2 \rangle - \langle H_1 \rangle = \frac{\hbar}{2}[-\tilde{\omega}_2 \tanh(\beta_H \hbar \tilde{\omega}_2) + \tilde{\omega}_1 \tanh(\beta_H \hbar \tilde{\omega}_1)];\\
&\beta_H = 1 / k_B T_{high}; \quad \tilde{\omega}_1 = 2\mu_B B_z(1);\\
&\tilde{\omega}_2 = 2\mu_B B_z(2); \quad \tilde{\omega}_1 > \tilde{\omega}_2.
\end{aligned}
\tag{14.18}
$$

The heat q_{out} released by the spin system is negative and given by:

$$
\begin{aligned}
&q_{out} = \langle H_4 \rangle - \langle H_3 \rangle = \frac{\hbar}{2}[-\tilde{\omega}_4 \tanh(\beta_C \hbar \tilde{\omega}_4) + \tilde{\omega}_3 \tanh(\beta_C \hbar \tilde{\omega}_3)];\\
&\beta_C = 1 / k_B T_{low}; \quad \tilde{\omega}_3 = 2\mu_B B_z(3); \quad \tilde{\omega}_4 = 2\mu_B B_z(4); \quad \tilde{\omega}_4 > \tilde{\omega}_3.
\end{aligned}
\tag{14.19}
$$

We find from the above equations that q_{in} is positive, while q_{out} is negative.

Since the adiabatic transitions between states 2 and 3, and between states 4 and 1, are adiabatic and isentropic, the entropy remains constant in our system. As a result, the probabilities P_1 and P_2 are also unchanged (as $P_1 + P_2 = 1$ in our system). It follows from Eq. (14.19) that the expectation value $\langle s_z \rangle = -(1/2)\tanh(\hbar\omega\beta/2)$ is also constant during these transitions. Using these relations, we get:

$$\beta_H \tilde{\omega}_1 - \beta_C \tilde{\omega}_4 \quad ; \quad \beta_H \tilde{\omega}_2 = \beta_C \tilde{\omega}_3 \quad ; \quad \frac{\tilde{\omega}_4}{\tilde{\omega}_1} = \frac{\tilde{\omega}_3}{\tilde{\omega}_2} = \frac{\beta_H}{\beta_C} = \frac{T_{low}}{T_{high}} . \tag{14.20}$$

Substituting Eq. (14.20) into Eqs. (14.18) and (14.19), we derive the efficiency of the Carnot reversible cycle as:

$$\eta_{rev} = \frac{q_{in} + q_{out}}{q_{in}} = 1 - \frac{T_{low}}{T_{high}} . \tag{14.21}$$

14.8. Case B: Irreversible Carnot cycle

Various studies have addressed the irreversibility of the Carnot cycle. We refer here to Esposito et al. (2010), where a simple model (without detailed treatment of the underlying interactions) was developed, providing reasonable agreement with experimental results under different conditions. We use here the idea that the irreversible heat exchange from the cold (or hot) reservoir will decay with time τ_C (τ_H) as:

$$Q_{ir,C} = T_{low}\left(-\Delta S - \frac{C_1}{\tau_C}\right); \quad Q_{ir,H} = T_{high}\left(\Delta S - \frac{C_2}{\tau_H}\right) . \tag{14.22}$$

Here, ΔS represents the entropy difference $S_h - S_l$, the subscripts H and C refer to the hot and cold reservoirs, respectively. C_1 and C_2 are constants, τ_C and τ_H are decay times, and the subscript ir represents irreversibility (the reversible regime is approached in the time limits

$\tau_C \to \infty$, $\tau_H \to \infty$). We consider the power generated during the Carnot cycle. Using Eq. (14.22), we get:

$$P = \frac{-W_{out}}{\tau_H + \tau_C} = \frac{(T_{high} - T_{low})\Delta S - T_{low}(C_1 / \tau_C) - T_{high}(C_2 / \tau_H)}{\tau_H + \tau_C} \quad . \qquad (14.23)$$

For $-W_{out}$, described in Fig. 1, the decay times are inserted as given in Eq. (14.23). The maximum power is found by setting the derivatives of P with respect to τ_H and τ_C equal to zero. Such a procedure was developed in Esposito et al. (2010), obtaining different results, including the case $C_1 = C_2$ for which Curzon-Ahlborn efficiency was obtained (Curzon and Ahlborn 1975). By neglecting the decay constants in Eq. (14.22) and using a relation similar to Eq. (14.21), we can derive in a simple way the efficiency of the reversible Carnot cycle.

14.9. Geometric phases applied to thermal spin-1/2 states.

Parallel transport equations for pure states are generalized to mixed states, by which the dynamical phases are eliminated. The geometric phase are then derived by unitary SU(2) transformations, including time-dependent parameters that are functions of the magnetic field interactions. A special form of the unitary transformation for mixed thermal states is obtained, through which geometric phases are acquired.

There are many works treating the effects of energy losses on the efficiency of quantum engines. However, our interest here is the possibility of using geometric phases in quantum engines. Since the dynamical phases obtained above for the Carnot engine (and similar dynamical phases for other engines) are based on expectation values of the Hamiltonian, they are not suitable for interference experiments. We present here an alternative approach for obtaining geometrical phases in thermal states, which can be applied in interferometers, e.g., in a Mach-Zehnder interferometer (Sjöqvist et al. 2000). The geometric phases obtained for thermal states are different from those obtained in NMR and neutron interferometry (Du et al. 2003; Ericsson et al. 2005; Ghosh and Kumar 2006; Klepp et al. 2005).

The density matrix ρ_0 for mixed states includes thermal fluctuations, which are described by the probabilities P_k associated with the states $|k\rangle_t$.

Therefore, unitary transformations cannot operate directly on the density matrix. Instead, they can operate separately on the components of the density matrix composed of pure quantum states. The idea presented in (Sjöqvist et al. 2000) is that the initial density matrix ρ_0 evolves in time t to a final density matrix ρ_t, by the unitary transformation operating on the initial vectors $|\tilde{k}\rangle_0$:

$$\rho_t = \sum_{k=1}^{N} P_k \left\{ U(t)\left|\tilde{k}\right\rangle_0 \left\langle \tilde{k}\right|_0 U^{\dagger}(t) \right\} = \sum_{k=1}^{N} P_k \left|\tilde{k}\right\rangle_t \left\langle \tilde{k}\right|_t, \tag{14.24}$$

where we use the short notations $|\tilde{1}\rangle_0$, $|\tilde{2}\rangle_0$, etc., for the initial states, $U(t)$ is a unitary matrix of $N \times N$ dimension, $|\tilde{k}\rangle_t = U(t)|\tilde{k}\rangle_0$ are the orthonormal vectors at time t, and $|\tilde{k}\rangle_0$ are the initial vectors. The mixed state is described as an incoherent mixture of several pure states, each weighted by its respective probability P_k. For thermal states, the eigenvectors in the computational basis are given by:

$$\left|\tilde{1}\right\rangle_0 = \begin{pmatrix} 1 \\ 0 \end{pmatrix} ; \left|\tilde{2}\right\rangle_0 = \begin{pmatrix} 0 \\ 1 \end{pmatrix}. \tag{14.25}$$

These correspond to eigenvalues -1/2 and +1/2 of $\hat{s}_z$. The density matrix of Eq. (14.14) can then be expressed in the short notation form:

$$\rho_0 = P_1 |-1/2\rangle\langle -1/2| + P_2 |+1/2\rangle\langle +1/2| \equiv P_1 \left|\tilde{1}\right\rangle\left\langle \tilde{1}\right| + P_2 \left|\tilde{2}\right\rangle\left\langle \tilde{2}\right|. \tag{14.26}$$

The parallel transport of a particular vector $|\tilde{k}\rangle_t$ implies that there is no change in phase when $|\tilde{k}\rangle_t$ evolves into $|\tilde{k}\rangle_{t+dt}$, under an infinitesimal change of the parameter t. Although there is no local phase change, the system can acquire a global geometric phase after completing a closed loop parametrized by t, or by Pancharatnam phase in the case of an open cycle. This phase is related to the curvature of the parameter space, depending only on the path's geometry. The condition for parallel transport can be satisfied by choosing a suitable $U(t)$ matrix which satisfies the parallel transport conditions from (Sjöqvist et al. 2000; Wagh and Rakheca 1995):

$$\langle k(t)|\frac{dU}{dt}U^{\dagger}|k(t)\rangle = 0 \quad ; \quad k = 1,2,\cdots,N \tag{14.27}$$

The geometric phase γ_{geom} is acquired by a mixed state evolving along a curve Γ under a unitary transformation, which satisfies the parallel transport conditions. These are given by (Sjöqvist et al. 2000):

$$\gamma_{geom}(\Gamma) = \arg Tr[\rho_t U(t)]$$
$$= \arg Tr[U(t)\rho_0 U^{\dagger}(t)U(t)] = \arg Tr[U(t)\rho_0]. \tag{14.28}$$

Here, the $U(t)$ for thermal states is given by a SU(2) two-dimensional unitary matrix. We choose $U(t)$ to be the product of 3 unitary matrices $U(t) = U_1(\tilde{\omega}_1)U_2(\xi)U_3(\tilde{\omega}_2)$:

$$U_1(\tilde{\omega}_1) = \begin{bmatrix} \exp[i\tilde{\omega}_1 t/2+\phi] & 0 \\ 0 & \exp[-i\tilde{\omega}_1 t/2+\phi] \end{bmatrix};$$
$$U_3(\tilde{\omega}_2) = \begin{bmatrix} \exp[i\tilde{\omega}_2 t/2] & 0 \\ 0 & \exp[-i\tilde{\omega}_2 t/2] \end{bmatrix}, \tag{14.29}$$

and

$$U_2(\xi) = \begin{bmatrix} \cos\xi & -\sin\xi \\ \sin\xi & \cos\xi \end{bmatrix}. \tag{14.30}$$

We assume that $U_1(\tilde{\omega}_1)$ and $U_3(\tilde{\omega}_2)$ operate during time t with Hamiltonians defined by Eq. (14.13) as $H_{1z} = \hbar\tilde{\omega}_1\sigma_z/2$ and $H_{2z} = \hbar\tilde{\omega}_2\sigma_z/2$ respectively. Here $\tilde{\omega}_1 = 2\mu_B B_{1z}$, $\tilde{\omega}_2 = 2\mu B_{2z}$, and B_{1z} and B_{2z} are constant magnetic fields in the *z*-direction. ϕ is a certain phase shift between $U_1(\tilde{\omega}_1)$ and $U_3(\tilde{\omega}_2)$ and $U_2(\xi)$ is a simple beam-splitter transformation which does not include any time dependence and, for simplicity, is defined by Eq. (14.30).

By substituting Eqs. (14.29) and (14.30) into Eq. (14.28), we get:

$$U(t) = e^{i\phi} \begin{pmatrix} e^{i\delta t} \cos\xi & -e^{-i\varsigma t} \sin\xi \\ e^{i\varsigma t} \sin\xi & e^{-i\delta t} \cos\xi \end{pmatrix}, \tag{14.31}$$

and

$$\delta = \frac{\tilde{\omega}_1 + \tilde{\omega}_2}{2} \quad ; \quad \varsigma = \frac{\tilde{\omega}_2 - \tilde{\omega}_1}{2}. \tag{14.32}$$

Eq. (14.31) is analogous to the SU(2) transformation used in neutron interferometry (Klepp et al. 2005). However, in the present case, the parameters δ and ς are multiplied by time *t*, resulting in a specific time dependence of *U*, which is critical for subsequent analysis.

The realization of the above unitary transformation depends on the relevant time scales. The initial mixed state, described by Eqs. (14.14) and (14.15) can be produced after a relatively long time, allowing the system to reach thermal equilibrium. The unitary transformation is then applied to the initial state during time t which is short relative to the relaxation times of the system. We use the above unitary matrix $U(t)$ for obtaining the parallel transport equations. Due to the linear relations between $|\tilde{k}\rangle_t$ and $|\tilde{k}\rangle_0$, the two parallel transport equations $\left\langle \tilde{k} \right|_t \frac{dU(t)}{dt} U^\dagger(t) \left| \tilde{k} \right\rangle_t = 0 \quad (k=1,2)$ also imply:

$$\left\langle \tilde{k} \right|_0 \frac{dU(t)}{dt} U^\dagger(t) \left| \tilde{k} \right\rangle_0 = 0 \quad ; \quad k = 1,2. \tag{14.33}$$

We use the eigenvectors of $|\tilde{k}\rangle_0$ in the computational basis, given by Eqs. (14.25). Using the first equation of (14.33), and for $U(t)$ from Eq. (14.31), we obtain:

$$(1,0) \begin{pmatrix} (i\delta\cos\xi) e^{i\delta t} & (i\varsigma \sin\xi) e^{-i\varsigma t} \\ (i\varsigma \sin\xi) e^{i\varsigma t} & (-i\delta\cos\xi) e^{-i\delta t} \end{pmatrix}$$
$$\times \begin{pmatrix} \cos\xi e^{-i\delta t} & \sin\xi e^{-i\varsigma t} \\ -\sin\xi e^{i\varsigma t} & \cos\xi e^{i\delta t} \end{pmatrix} \begin{pmatrix} 1 \\ 0 \end{pmatrix} = 0. \tag{14.34}$$

Straightforward calculations lead to the result:

$$((i\delta\cos\xi)e^{i\delta t} \quad (i\varsigma\sin\xi)e^{-i\varsigma t})$$
$$\times\begin{pmatrix}\cos\xi e^{-i\delta t}\\ -\sin\xi e^{i\varsigma t}\end{pmatrix} = i\delta(\cos\xi)^2 - i\varsigma(\sin\xi)^2 = 0. \tag{14.35}$$

For the second equation of (14.33), we apply the same calculations from equation (14.34), in which we substitute $(1 \quad 0)$ to $(0 \quad 1)$ and $\begin{pmatrix}1\\0\end{pmatrix}$ to $\begin{pmatrix}0\\1\end{pmatrix}$. Then we get:

$$(i\varsigma\sin\xi e^{i\varsigma t} \qquad -i\delta\cos\xi e^{-i\delta t})$$
$$\times\begin{pmatrix}e^{-i\varsigma t}\sin\xi\\ e^{i\delta t}\cos\xi\end{pmatrix} = i\varsigma(\sin\xi)^2 - i\delta(\cos\xi)^2 = 0. \tag{14.36}$$

The equivalent Eqs. (14.35) and (14.36) lead to a sufficient condition for parallel transport:

$$(\tan\xi)^2 = \frac{\delta}{\varsigma} \quad ; \quad \cos^2\xi\left(1+\frac{\delta^2}{\varsigma^2}\right) = 1 \rightarrow \cos\xi = \sqrt{\frac{\varsigma^2}{\delta^2+\varsigma^2}}. \tag{14.37}$$

Using Eq. (14.25) and Eq. (14.28) for $U(t)$, we get:

$$\langle 1|_0 U(t)|1\rangle_0 = \cos\xi e^{i(\delta t+\phi)} \quad ; \quad \langle 2|_0 U(t)|2\rangle_0 = \cos\xi e^{-i(\delta t+\phi)}. \tag{14.38}$$

The geometric phase γ_{geom} is obtained by the incoherent superposition of two fields produced separately from the two components of the thermal field. Therefore, it is given by:

$$\gamma_{geom} = \arg\left[P_1\cos\xi e^{i(\delta t+\phi)} + P_2\cos\xi e^{-i(\delta t+\phi)}\right]. \tag{14.39}$$

From Eq. (14.39), we get the result:

$$\gamma_{geom} = \arctan\left[\left(\frac{P_1-P_2}{P_1+P_2}\right)\tan(\delta t+\phi)\right], \tag{14.40}$$

where, for the present system $P_1 + P_2 = 1$. The radiation intensity is reduced by the factor:

$$\tilde{v} = | P_1 \cos\xi e^{i(\delta t+\phi)} + P_2 \cos\xi e^{-i(\delta t+\phi)} | \\ = \cos\xi \sqrt{P_1^2 + P_2^2 + 2P_1P_2 \cos(2\delta t + 2\phi)} \, . \quad (14.41)$$

The unitary development of the spin-1/2 system was produced by the system's consecutive interactions with magnetic fields, which were applied over time scales that were short relative to the system's relaxation times. Parallel transport equations, which eliminate the dynamical phase, were obtained by assuming the parameter relations specified in Eq. (14.37). For the present mixed spin-1/2 system $\delta t + \phi$ represents Pancharatnam's phase for an open cycle, and $(P_1 - P_2)/(P_1 + P)$ represents the thermal noise parameter.

Using equation (14.15), the thermal noise term can be transformed to:

$$\left(\frac{P_1 - P_2}{P_1 + P_2} \right) = \tanh\left[\frac{\hbar\tilde{\omega}\beta}{2} \right] . \quad (14.42)$$

For very low temperatures, this term tends to the value 1. For very high temperatures, we can use the approximations:

$$\sinh\left[\frac{\hbar\tilde{\omega}\beta}{2} \right] = \frac{\hbar\tilde{\omega}\beta}{2} \quad ; \quad \cosh\left[\frac{\hbar\tilde{\omega}\beta}{2} \right] = 1 \quad ; \quad \left(\frac{P_1 - P_2}{P_1 + P_2} \right) = \frac{\hbar\tilde{\omega}}{2k_B T} , \quad (14.43)$$

so that the term is given by the ratio between the quantum vacuum energy $(\hbar\tilde{\omega})/2$ and the thermal energy $k_B T$. The geometric phase in this case is obtained as a Pancharatnam phase for an open cycle. We note that the Pancharatnam phase for open cycles was also observed in neutron experiments (Klepp et al. 2005).

14.10. Discussion

The efficiency of thermal engines and the role of time durations in achieving maximum efficiency were analyzed for both the Otto harmonic oscillator engine and the Carnot engine using a spin-1/2 system. A two-dimensional Carnot thermal spin-1/2 system with time-dependent magnetic fields in the symmetric *z*-direction was analyzed. The density matrix ρ_0, obtained for mixed states, includes thermal fluctuations averaged via the probability $P_k\,(k=1,2)$, and therefore, cannot be directly used in interferometry. However, the application of unitary transformation of mixed states, including the parallel transport equations, was found to be of special interest, as these methods eliminate dynamical phases, leaving only geometric phases. We find that the geometric phases obtained through the unitary transformation are Pancharatnam's phases, which are gauge-invariant and valid also for open cycles. It is interesting to note that geometric phases can be obtained in quantum engine systems, giving rise to interference effects that are fundamentally different from those commonly observed in nuclear magnetic resonance and neutron interferometry.

In conclusion, geometric phases obtained in quantum engines, although not yet experimentally observed, may lead to new insights into the field of quantum thermodynamics.

References

Abah, O., J. Robnagel, G. Jacob, S. Deffner, F. Schmidt-Kaler, K. Singer, and E. Lutz. 2012. "Single-Ion Heat Engine at Maximum Power." *Physical Review Letters* 109: 203006. https://doi.org/10.1103/PhysRevLett.109.203006.

Ben-Aryeh, Y. 2004. "Berry and Pancharatnam Topological Phases of Atomic and Optical Systems." *Journal of Optics B: Quantum and Semiclassical Optics* 6: R1–R18. https://doi.org/10.1088/1464-4266/6/4/R01.

Ben-Aryeh, Y. 2025. "Quantum Thermal Engine Operating with Spin 1/2 System and Geometric Phases Obtained by Unitary Transformation Matrix of Mixed States." *International Journal of Thermodynamics* Erken Görünüm-Early Pub Issues: 1-6. doi: 10.5541/ijot.1395683.

Ben-Aryeh, Y. 2022. "Quantum Dynamics and Geometric Phases in Harmonic Oscillator, Spin ½ and Two-Level Thermal Engine Systems." https://doi.org/10.48550/arXiv.2212.02970.

Berry, M. V. 1984. "Quantal Phase Factors Accompanying Adiabatic Changes." *Proceedings of the Royal Society of London*. Series A 392: 45. https://doi.org/10.1098/rspa.1984.0023.

Curzon, F. L., and B. Ahlborn. 1975. "Efficiency of a Carnot Engine at Maximum Power output." *American Journal of Physics* 43: 22. https://doi.org/10.1119/1.10023.

Du, J., P. Zou, M. Shi, L. C. Kwek, J.-W. Pan, C. H. Oh, A. Ekert, D. K. L. Oi, and M. Ericsson. 2003. "Observation of Geometric Phases for Mixed States Using NMR Interferometry." *Physical Review Letters* 91: 100403. https://doi.org/10.1103/PhysRevLett.91.100403.

Ericsson, M., D. Achilles, J. T. Barreiro, D. Branning, N. A. Peters, and P. G. Kwiat. 2005. "Measurement of Geometric Phase for Mixed State Using Single Photon Interferometry." *Physical Review Letters* 94: 050401. https://doi.org/10.1103/PhysRevLett.94.050401.

Esposito, M., R. Kawai, K. Lindenberg, and C. Van den Broeck. 2010. "Efficiency at Maximum Power of Low-Dissipation Carnot Engines." *Physical Review Letters* 105: 150603. https://doi.org/10.1103/PhysRevLett.105.150603.

Geva, E., and R. Kosloff. 1992. "A Quantum-Mechanical Heat Engine Operating in Finite Time: A Model Consisting of Spin-1/2 Systems as the Working Fluid." *Journal of Chemical Physics* 96: 3054. https://doi.org/10.1063/1.461951.

Ghosh, A., and A. Kumar. 2006. "Experimental Measurement of Mixed State Geometric Phase by Quantum Interferometry Using NMR." *Physics Letters A* 349: 27. https://doi.org/10.1016/j.physleta.2005.08.092.

Jarzynski, C. 1997. "Nonequilibrium Equality for Free Energy Differences." *Physical Review Letters* 78: 2690–2693. https://doi.org/10.1103/PhysRevLett.78.2690.

Klepp, J., S. Sponar, Y. Hasegawa, E. Jericha, and G. Badurek. 2005. "Noncyclic Pancharatnam Phase for Mixed SU(2) Evolution in Neutron Polarimetry." *Physics Letters* A 342: 48. https://doi.org/10.1016/j.physleta.2005.05.038.

Kosloff, R., and Y. Rezek. 2017. "The Quantum Harmonic Otto Cycle." Entropy 19: 136. https://doi.org/10.3390/e19040136.

Leff, H. S. 1987. "Thermal Efficiency at Maximum Work Output: New Results for Old Heat Engines." *American Journal of Physics* 55: 602. https://doi.org/10.1119/1.15071.

Lin, B., and J. Chen. 2005. "Optimal Analysis of the Performance of an Irreversible Quantum Heat Engine with Spin Systems." *Journal of Physics A: Mathematical and General* 38: 69. https://doi.org/10.1088/0305-4470/38/1/004.

Robnagel, J., S. T. Dawkins, K. N. Tolazzi, O. Abah, E. Lutz, F. Schmidt-Kaler, and K. Singer. 2016. "A Single-Atom Heat Engine." *Science* 352: 325. https://doi.org/10.1126/science.aad6320.

Sjöqvist, E., A. K. Pati, A. Ekert, J. S. Anandan, M. Ericsson, D. K. L. Oi, and V. Vedral. 2000. "Geometric Phases for Mixed States in Interferometry." *Physical Review Letters* 85: 2845. https://doi.org/10.1103/PhysRevLett.85.2845.

Wagh, A. G., and V. C. Rakhecha. 1995. "On Measuring the Pancharatnam Phase. I. Interferometry." *Physics Letters A* 197: 107. https://doi.org/10.1016/0375-9601(94)00914-B.

PART III

SUPER-RESOLUTION IMAGING RELYING ON LIGHT NANOSCALE SCATTERING

CHAPTER 15

HOT SPOTS IN TWO SYMMETRIC METALLIC SPHERES OBTAINED BY LAPLACE EQUATION SOLUTIONS USING BI-SPHERICAL COORDINATES

15.1. Introduction

Raman signals in molecules produced by incident electro-magnetic (EM) fields are strongly amplified when the molecules are located in the gaps between the nanoparticles, due to the intense EM fields induced in these regions (known as “hot spots”) (Le Ru and Etchegoin 2009; Ding et al. 2017). Experimental results on surface-enhanced Raman scattering (SERS) have been interpreted through the hot spot mechanism (Inoue and Ohtaka 1983; Jiang et al. 2003; Nikoobakht and El-Sayed 2003). Notably, extensive studies have focused on Raman scattering in nanoparticle dimers (Lee et al. 2013; Li et al. 2008). It was found that the signal in SERS is proportional to the fourth power of the amplified EM field when the incident polarization is aligned parallel to the interparticle axis of the dimer (i.e., under parallel polarization). Enhancement of EM fields through their interactions with local metallic surface plasmons is well known. Various methods have been developed to analyze hot spot phenomena, including discrete dipole approximation (Yang, Schatz, and Van Duyne 1995; Hao and Schatz 2004). Building on such approaches, the optical modeling techniques rely on solving Maxwell’s equations for each specific geometry and illumination conditions. These methods were based on adaptive grids that assign fine discretization to the EM fields. Using this framework, the Van Dyne group has conducted detailed analyses of hot spot behavior, including the influence of *retardation* effects (Kleinman et al. 2013). When the dimensions of the optical system are very small relative to the wavelength, retardation effects can be neglected. In this regime, the magnetic field is negligible at optical frequencies, and the electric field is given as $\vec{E} = -\vec{\nabla}\psi$, where ψ is the solution of the

Laplace equation. In the present work, we apply these conditions to treat hot spots in two metallic spheres using bi-spherical coordinates (Morse and Feshbach 1953; Aravind, Nitzan, and Metiu 1981; Stoy 1989a; Stoy 1989b; Ruppin 1982; Goyette and Navon 1976; Ben-Aryeh 2021; Ben-Aryeh 2023). Although we treat a special system, its solutions provide insights into the general properties of hot spots.

15.2. Applications of bi-spherical coordinates

Figure 15.1 illustrates the bi-spherical coordinates η, α of the dimer as a function of the $x-z$ coordinates, where, for simplicity, the third bi-spherical coordinate is assumed to be $\phi=0$. The dimer potential does not change under rotation around the *z*-axis (changing ϕ) so that the diagram is described in the $x-z$ plane for a certain value of the parameter a. Thus, we have the relations:

$$x = a\sin\alpha / (\cosh\eta - \cos\alpha);$$
$$z = a\sinh\eta / (\cosh\eta - \cos\alpha). \tag{15.1}$$

The diagram describes spheres whose cross sections in the $x-z$ plane are circles (the large circles are truncated from the figure), and each circle corresponds to a certain η value. On the right-hand side of the figure, the circles correspond to $\eta = 0.2,\ 0.4,\ 0.8,\ \text{and}\ 1.3$, and on the left side to $\eta = -0.2,\ -0.4,\ -0.8,\ \text{and}\ -1.3$. For a certain a value the radius R of each sphere and the distance D between the sphere center and the origin of the coordinate system are given, respectively, as:

$$R = \frac{a}{\sinh\eta} \quad ; \quad D = a\coth\eta \quad . \tag{15.2}$$

The diagram shows the poles for which $\eta \to \pm\infty$, $R \to 0$, and $D \to a$, where the distance between the two poles is $2a$. Surfaces of bi-spherical spheres satisfy the relations:

$$x^2 + y^2 + (z - a\coth\eta)^2 = \frac{a^2}{\sinh^2\eta}, \tag{15.3}$$

where $a/\sinh\eta$ is the radius *R* of the sphere, and the center of the sphere is shifted from the origin along the z-axis by a distance $D = a\coth\eta$. The shortest distance from the origin to the spherical surface is given by:

$$\delta = a\coth\eta - \frac{a}{\sinh\eta}, \tag{15.4}$$

so that the width of the hot spot is 2δ.

Curves of the polar angle α are orthogonal to the curves of constant η, and are described in Figure 15.1 for $\alpha = \pi,\ \pi/2,\ \pi/3, \text{ and } \pi/6$. The homogeneous EM fields, polarized in the z-direction, propagate along the *x*-direction. In this chapter, we restrict the analysis to parallel polarization, where the incident polarization is parallel to the inter-particle axis of the dimer, and the polarization charges face each other across the narrow gap.

Bi-spherical coordinates in the $x-z$ plane, are given by circles of constant bi-spherical radial coordinate $\eta = \pm 0.2,\ 0.4,\ 0.8, \text{ and } 1.3$, and curves of constant polar angle are given for $\alpha = \pi,\ \pi/2,\ \pi/3, \text{ and } \pi/6$. The EM field, polarized in the z-direction is propagating in the x-direction. The curve for $\alpha = \pi$ is along the z-axis and the poles with $\eta = \pm\infty$ are located on the z-axis, separated by a distance $2a$.

We restrict the analysis to a symmetric dimer, as bi-spherical coordinates significantly simplify the description of the metallic surfaces and the application of boundary conditions. In cases where the two spheres have different radii, the use of bi-spherical coordinates becomes considerably more complicated.

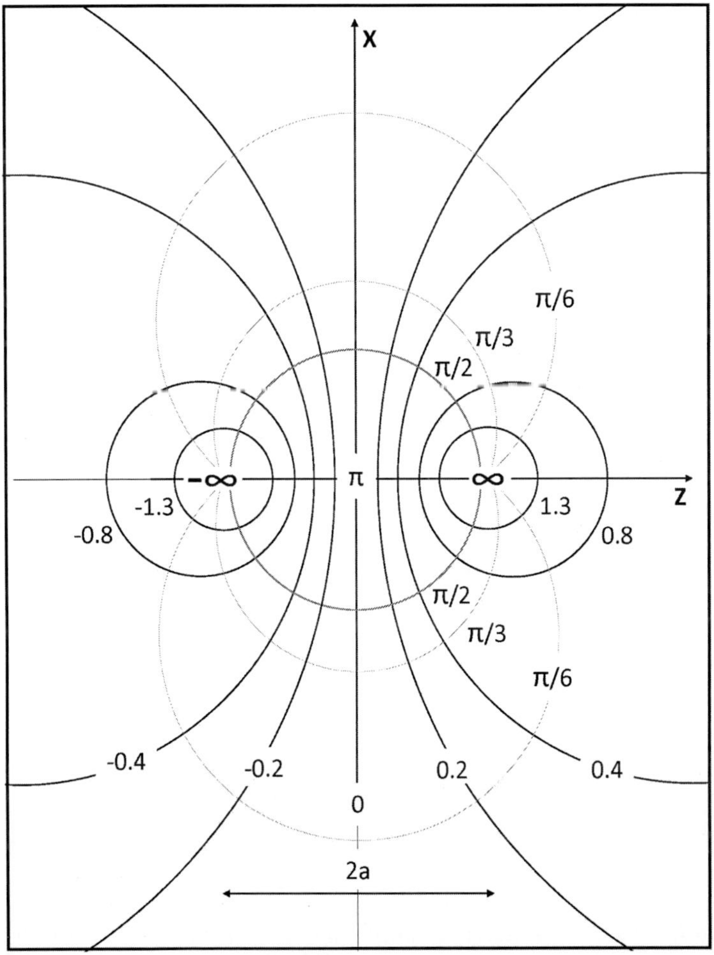

Figure 15.1: Bi-spherical coordinates for two symmetric metallic spheres. Reproduced from Ben-Aryeh, Y. 2023. "Analytical Results for Enhancement Factor (EF) of SERS for Two Metallic Spheres and Nano-Shells." *AIP Advances* 13: 035236. Licensed under CC BY.

In order to apply bi-spherical coordinates to a specific dimer, for which *the radius* R *and the shift* D (the displacement of the sphere center from the origin) *are fixed by the experimental conditions*, we need to use the inverse conditions of Eqs. (15.2):

$$\cosh\eta_0 = (D/R)_{\exp} \quad ; \quad a_0 = \left[R\sqrt{(D^2/R^2)-1}\right]_{\exp} \; ;$$

$$\sinh\eta_0 = \left[\sqrt{(D^2/R^2)-1}\right]_{\exp} . \tag{15.5}$$

Eq. (15.5) represents a particular dimer with circular cross sections, corresponding in Figure 15.1 to specific values of $\pm\eta_0$.

15.3. Solutions with bi-spherical coordinates, for the potential and the electric field at the hot spot of two symmetric metallic spheres, using boundary conditions with approximations.

Solutions to the Laplace equation for two symmetric metallic spheres under an incident EM field parallel to the symmetric *z*-axis were analyzed by Stoy (1989a) as:

A. Outside the two spheres:

$$\psi^{(1)}(\eta,\alpha) = (\cosh\eta - \cos\alpha)^{1/2};$$

$$\sum_{n=0}^{\infty} P_n(\cos\alpha)(\cosh\eta - \cos\alpha)^{1/2}$$

$$\left[A_n \sinh\left(n+\frac{1}{2}\right)\eta - E_0 a_0 2^{1/2}(2n+1)\exp\left[-\eta(n+1/2)\right]\right], \tag{15.6}$$

where $P_n(\cos\alpha)$ are Legendre polynomials, A_n are the coefficients corresponding to the solution outside the two spheres. The sum of terms proportional to $-E_0$ represent the external potential $-E_0 z$ of the EM field, where z is expanded into n terms (Stoy 1989a).

B. Inside the two spheres:

$$\psi^{(2)}(\eta,\alpha) = (\cosh\eta - \cos\alpha)^{1/2}\sum_{n=0}^{\infty} P_n(\cos\alpha) B_n \exp\left[-\eta(n+1/2)\right] . \tag{15.7}$$

This solution decays to zero at the poles, where $\eta = \pm\infty$, and B_n are certain coefficients. The above solutions were obtained by using the symmetries of these systems (Stoy 1989a; Stoy 1989b).

The above equations can be simplified in the region of the hot spot, i.e., in the narrow gap between the two spheres, for which we use the approximation:

$$P_n(\cos\alpha) = P_n(\cos\pi) = P_n(-1) = \begin{cases} 1 \textit{ for even number } n \\ -1 \textit{ for odd number } n \end{cases}. \tag{15.8}$$

We assume that the surfaces of the two spheres are obtained by the bi-spherical coordinates $\pm\eta_0$. According to the first boundary condition:

$$\left[\psi^{(1)}\right]_{\eta=\eta_0} = \left[\psi^{(2)}\right]_{\eta=\eta_0} \Rightarrow B_n \exp\left[-\eta(n+1/2)\right]$$
$$= A_n \sinh\left[(n+1/2)\eta_0 - E_0 a_0 2^{1/2}(2n+1)\exp[-\eta_0(n+1/2)]\right]. \tag{15.9}$$

According to the second boundary condition:

$$\varepsilon_1\left[\frac{\partial}{\partial\eta}\psi_1\right]_{\eta=\eta_0} = \varepsilon(\omega)\left[\frac{\partial}{\partial\eta}\psi_2\right]_{\eta=\eta_0}, \tag{15.10}$$

where $\varepsilon(\omega)$ and ε_1 are the dielectric constants of the metallic spheres and the surrounding medium, respectively. At the hot spots, the derivatives of the terms proportional to A_n and B_n are very large, as they represent the amplified EM field, which is larger than the external field $-E_0 z$, and are proportional to high values of n. These dominant derivatives allow the neglect of all other derivative terms. Applying this approximation to Eq. (15.10) gives:

$$\varepsilon_1 A_n\left[\cosh(n+1/2)\eta_0\right] = -\varepsilon(\omega) B_n \exp[-(n+1/2)\eta_0]. \tag{15.11}$$

The approximation introduced in Eq. (15.11) leads to separate equations for each integer n, and is valid in the small gap region, where the EM fields are extremely large. By comparing Eq. (15.9), multiplied by $\varepsilon(\omega)$ with Eq. (15.11), we get:

$$\varepsilon(\omega)\begin{pmatrix} A_n \sinh\left[(n+1/2)\eta_0\right] \\ -E_0 a_0 2^{1/2}(2n+1) \\ \exp\left[-\eta_0(n+1/2)\right] \end{pmatrix} + \varepsilon_1\left(A_n \cosh(n+1/2)\eta_0\right) = 0. \tag{15.12}$$

The equation for A_n is then given by:

$$A_n = \frac{2^{1/2}(2n+1)\varepsilon(\omega)E_0 a_0 \exp\left[-(n+1/2)\eta_0\right]}{\varepsilon_1 \cosh\left[(n+1/2)\eta_0\right] + \varepsilon(\omega)\sinh\left[(n+1/2)\eta_0\right]}. \tag{15.13}$$

By transforming the hyperbolic functions to exponential terms and dividing both the numerator and the denominator by $\exp[-(n+1/2)\eta_0]$, we get:

$$A_n = \frac{2^{1/2}(2n+1)\varepsilon(\omega)E_0 a_0}{\left[\varepsilon_1 + \varepsilon(\omega)\right]\exp\left[(2n+1)\eta_o\right] + \left[\varepsilon_1 - \varepsilon(\omega)\right]}. \tag{15.14}$$

By substituting $P_n(\cos\alpha)$ according to Eq. (15.8), into Eq. (15.6), and neglecting the terms proportional to E_0 in Eq. (15.6), we get the final equation for the potential in the hot spot of the two symmetric metallic spheres:

$$\psi^{(1)}(\eta) = (\cosh\eta - \cos\alpha)^{1/2}\sum_{n=0}^{\infty}(-1)^n A_n \sinh\left[\left(n+\frac{1}{2}\right)\eta\right], \tag{15.15}$$

Where the coefficients A_n are given by Eq. (15.14). The corresponding EM field in the hot spot is given by (Stoy 1989a):

$$E_{spot} = -\left[(\cosh\eta - \cos\alpha)/a\right]\frac{d}{dt}\psi^{(1)}. \tag{15.16}$$

Using similar approximations to those made above, we obtain the expression for the EM field in the hot spot of the two metallic spheres:

$$E_{spot} = \sum_{n-0}^{\infty} E_n \;;$$

$$E_n = -\left[\left(\cosh\eta - \cos\alpha\right)^{3/2} / a\right]$$

$$\left[(n+1/2)\cosh\left[(n+1/2)\eta\right](-1)^n A_{n.}\right] \tag{15.17}$$

It should be noted that the amplified EM field in the hot spot is proportional to the external EM field. As a result, terms proportional to the square of the external EM field have been neglected. However, the proportionality constant between the amplified field and the external field in the hot spot can be represented by an extremely large number.

15.4. Surface-enhanced Raman scattering (SERS) at the center of the two symmetric metallic spheres

We are interested in the calculation of the total EM light intensity at the hot spot, given by:

$$\left\langle E_{spot}{}^{2}\right\rangle = \sum_{n-0}^{\infty} \left|E_n\right|^2 \;;\; \left\langle E_{spot}{}^{4}\right\rangle = \left[\sum_{n-0}^{\infty} \left|E_n\right|^2\right]^3 . \tag{15.18}$$

In the evaluation of $E_{spot}{}^{2}$, we note that the non-diagonal products $E_n E_{n'}\left(n \neq n'\right)$ with alternating signs canceling out, so their total contribution vanishes. Therefore, only the diagonal terms of $\left\langle E_{spot}{}^{2}\right\rangle$ are considered in Eq. (15.18). As the SERS is proportional to $\left\langle E_{spot}{}^{4}\right\rangle$, it is given by the square of the field intensity as given in Eq. (15.18). We also find it of interest to calculate the maximum amplification of the light intensity I / I_0 in the hot spot:

$$\left(I / I_0\right) = \frac{1}{E_0^{\,2}} \sum_{n-0}^{\infty} \left|E_n\right|^2 . \tag{15.19}$$

To obtain the hot spot, we assume that the distance between the two spheres is very small relative to their radius, and that the hot spot is located on the x-axis, where $\alpha \to \pi$. Following these conditions, we substitute:

$$\cosh\eta = 1 \ ; \ \cos\alpha = -1 \ ; \ \cosh\left[(n+1/2)\eta\right] \to 1 \ \ for \ \ \eta \ll 1 \tag{15.20}$$

into Eq. (15.17), and obtain:

$$|E_n| = \left(\frac{1}{a_0}\right)(n+1/2)2^{3/2} A_n \tag{15.21}$$

By substituting A_n from Eq. (15.14) into Eq. (15.21), we get:

$$|E_n| = \left(\frac{1}{a_0}\right)(n+1/2)2^{3/2} \frac{2^{1/2}(2n+1)\varepsilon(\omega)E_0 a_0}{\left[\varepsilon_1 + \varepsilon(\omega)\right]\exp\left[(2n+1)\eta_o\right] + \left[\varepsilon_1 - \varepsilon(\omega)\right]}. \tag{15.22}$$

We note that the parameter a_0 cancels out in Eq. (15.22), indicating that this equation represents the amplification of the light intensity. To proceed, we transform the summation into an integral using the following relations:

$$(n+1/2)2\eta_0 = x \ \ ; \ \ dx = 2\eta_0 dn \tag{15.23}$$

$$\int_0^\infty \left|\frac{E_n}{E_0}\right|^2 dn = \left[\varepsilon(\omega)/(\varepsilon_1 + \varepsilon(\omega))\right]^2 \frac{2}{(\eta_0)^5} \int_0^\infty \frac{x^4}{\left(e^x + G(\omega)\right)^2} dx \ ;$$
$$G(\omega) = \left[\varepsilon_1 - \varepsilon(\omega)\right] / \left[\varepsilon_1 + \varepsilon(\omega)\right]. \tag{15.24}$$

Eq. (15.24) can be evaluated numerically, where small values of $G(\omega)$ lead to a reduction in the intensity amplification factor. By assuming a very large negative value of $\varepsilon(\omega)$ we get the approximation $G(\omega) \to -1$. Under this approximation, the integral in Eq. (15.24) can be calculated analytically (Morse and Feshbach 1953), and the amplification of the light intensity is given by:

$$\int_0^\infty \left|\frac{E_n}{E_0}\right|^2 dn = \left[\frac{\varepsilon(\omega)}{\varepsilon_1+\varepsilon(\omega)}\right]^2 \frac{2.18}{\eta_0^5} . \tag{15.25}$$

Since the amplification in SERS measurements is proportional to the fourth power of the EM field, its enhancement factor (EF) is given by the square of Eq. (15.25):

$$EF = \left[\frac{\varepsilon(\omega)}{\varepsilon_1+\varepsilon(\omega)}\right]^4 \frac{4.76}{\eta_0^{10}} . \tag{15.26}$$

15.5. Summary, discussion, and conclusions

For treating hot spots in molecular systems, when their dimensions are smaller than the wavelength, Laplace equation solutions can be used, as magnetic fields are negligible in this regime. In our analysis, we focused on the specific case of two closely spaced, symmetric metallic spheres. The use of bi-spherical coordinates in this configuration greatly simplifies the application of boundary conditions. While the general analysis of this system is quite complex, several approximations can be applied to describe the extreme amplification of the EM field in the narrow gap between the two spheres (the hot spot):

1. The surfaces of the two spheres are defined as the bi-spherical coordinates $\pm\eta_0$, where R is the radius of the spheres, $\sinh\eta_0 = \sqrt{(D/R)^2-1}$, and D is the distance from the sphere's center to the origin of the coordinate system. The gap width $2D-2R=2\delta$, which defines the hot spot, is assumed to be very small.
2. The Laplace solutions for the space outside and inside the two spheres are given, respectively, in Eqs. (15.6) and (15.7), as summations over terms. These expressions are simplified in the region of the hot spot due to the condition $\alpha=\pi$ valid in this region (see Figure 15.1). Then, in the hot spot, the Legendre polynomials $P_n(\cos\alpha)$ reduce to 1 for even n, and to -1 for odd n.
3. For the second boundary condition, we get analytical results which include derivatives of the potentials on both sides of the metallic

> surfaces (at $\eta = \eta_0$). At the hot spots, the coefficients A_n and B_n corresponding to large n values become dominant, and their derivatives are very large relative to those of other terms. Using this approximation, the equations for different n values become decoupled. With the aid of the first boundary condition, we then derive analytical expressions for the coefficients A_n in Eq. (15.14), which is a function of the parameter η_0. In a more general analysis, the second boundary condition leads to coupling between equations with different n values. Therefore, the present approximation is valid only in the hot spot regime, where there is high amplification of the EM field.

The EM field at the hot spot was obtained from the bi-spherical derivative of the potential, as given in Eq. (15.16). Due to the alternating sign of the n'th component of the EM field as a function of n, the amplification of light in the hot spot is defined as $(I / I_0) = (1 / E_0{}^2)\sum_{n=0}^{\infty}|E_n|^2$, as expressed in Eq. (15.19), where only $|E_n|^2$ is well defined. By following the same approximations used earlier and transforming the summation into an integral, the final expression for the amplification of light intensity is given by the integral in Eq. (15.24). A key parameter in this equation is given by $G(\omega) = [\varepsilon_1 - \varepsilon(\omega)] / [\varepsilon_1 + \varepsilon(\omega)]$. For metals with a large imaginary component of $\varepsilon(\omega)$ (as in the case of Ag), this parameter tends to the value -1, resulting in very large values for light amplification as obtained in Eq. (15.25), which is proportional to $\eta_0{}^{-5}$. As the SERS factor is proportional to the fourth power of the EM field, it is described by Eq. (15.26), and is proportional to $\eta_0{}^{-10}$.

There are certain features of the present analysis for two symmetric metallic spheres that are common to other SERS measurements: a) A small gap is a necessary condition for obtaining a hot spot, i.e., for high amplification of the EM field. b) A large negative value of the real part of the dielectric constant $\varepsilon(\omega)$ is also a necessary condition for the existence of hot spots. These properties can be qualitatively explained by the idea that the EM in the hot spot does not penetrate the metal when it has a large negative value of $\varepsilon(\omega)$. The resulting reduction in the cross-section for the transmitted EM field guarantees an increase in light intensity within the gap. For extremely small gaps, much smaller than the wavelength, the light in the gap becomes almost electrostatic, and the magnetic field becomes negligible. Very large EF factors have been reported by various authors (e.g., Yan et al. 2021; Le Ru, Etchegoin, and Meyer 2006; McMahon et al. 2012), with properties consistent with those presented here for two symmetric metallic spheres.

References

Aravind, P. K., A. Nitzan, and H. Metiu. 1981. "The Interaction between Electromagnetic Resonances and Its Role in Spectroscopic Studies of Molecules Adsorbed on Colloidal Particles or Metal Spheres." *Surface Science* 110: 189. https://doi.org/10.1016/0039-6028(81)90595-1

Ben-Aryeh, Y. 2021. "Hot Spots in Two Metallic Spheres System Related to Laplace Equation Solutions with Bi-Spherical Coordinates." *Applied Physics B* 127: 127–157. https://doi.org/10.1007/s00340-021-07705-7

Ben-Aryeh, Y. 2023. "Analytical Results for Enhancement Factor (EF) of Surface Enhanced Raman Spectroscopy (SERS) for Two Metallic Spheres and Nano-Shells." *AIP Advances* 13: 035236. https://doi.org/10.1063/5.0141047

Ding, S. Y., E. M. You, Z.-Q. Tian, and M. Moskovits. 2017. "Electromagnetic Theories of Surface Enhanced Raman Spectroscopy." *Chemical Society Reviews* 46: 4042. https://doi.org/10.1039/C7CS00238F

Goyette, A., and A. Navon. 1976. "Two Dielectric Spheres in an Electric Field." *Physical Review B* 13: 4320. https://doi.org/10.1103/PhysRevB.13.4320

Hao, E., and G. C. Schatz. 2004. "Electromagnetic Fields around Silver Nanoparticles and Dimers." *Journal of Chemical Physics* 120: 357–366. https://doi.org/10.1063/1.1629280

Inoue, M., and K. Ohtaka. 1983. "Surface Enhanced Raman Scattering by Metal Spheres. I. Cluster Effect." *Journal of the Physical Society of Japan* 52: 3853. https://doi.org/10.1143/JPSJ.52.3853

Jiang, J., K. Bosnick, M. Maillard, and L. Brus. 2003. "Single Molecule Raman Spectroscopy at the Junctions of Large Ag Nanocrystals." *Journal of Physical Chemistry B* 107: 9964. https://doi.org/10.1021/jp034632u

Kleinman, S. L., R. R. Frontiera, A.-I. Henry, J. A. Dieringer, and R. P. Van Duyne. 2013. "Creating, Characterizing and Controlling

Chemistry with SERS Hot Spots." *Physical Chemistry Chemical Physics* 15: 21. https://doi.org/10.1039/C2CP42598J

Le Ru, E. C., and P. G. Etchegoin. 2009. "Phenomenology Local Field Enhancement Factor Distributions around Electromagnetic Hot Spots." *Journal of Chemical Physics* 130: 181101. https://doi.org/10.1063/1.3138784

Le Ru, E. C., P. G. Etchegoin, and M. Meyer. 2006. "Enhancement Factor Distribution around a Single Surface-Enhanced Raman Scattering Hot Spot and Its Relation to Single Molecule detection." *Journal of Chemical Physics* 125: 204701. https://doi.org/10.1063/1.2390694

Lee, H. M., J.-H. Lee, S. M. Jin, Y. D. Suh, and J.-M. Nam. 2013. "Single-Molecule and Single-Particle Based Correlation Studies between Localized Surface Plasmons of Dimeric Nanostructures with ~1 nm Gap and Surface Enhanced Raman Scattering." *Nano Letters* 13: 6113. https://doi.org/10.1021/nl4034297

Li, W., P. H. C. Camargo, X. Lu, and Y. Xia. 2008. "Dimers of Silver Nanospheres: Facile Synthesis and Their Use as Hot Spots for Surface Enhanced Raman Scattering." *Nano Letters* 9: 485. https://doi.org/10.1021/nl803621x

McMahon, J. M., S. Li, L. K. Ausman, and G. C. Schatz. 2012. "Modelling the Effects of Small Gaps in Surface-Enhanced Raman Spectroscopy." *Journal of Physical Chemistry C* 116: 1637. https://doi.org/10.1021/jp207661y

Morse, P. M., and H. Feshbach. 1953. *Methods of Theoretical Physics*. New York: McGraw-Hill.

Nikoobakht, B., and M. A. El-Sayed. 2003. "Surface Enhanced Raman Scattering Studies on Aggregated Gold Nanorods." *Journal of Physical Chemistry A* 107: 3372. https://doi.org/10.1021/jp026770+

Ruppin, R. 1982. "Surface Modes of Two Spheres." *Physical Review B* 26: 3440. https://doi.org/10.1103/PhysRevB.26.3440

Stoy, R. D. 1989a. "Solution Procedure for the Laplace Equation in Bi-Spherical Coordinates for Two Spheres in a Uniform External Field: Parallel Orientation." *Journal of Applied Physics* 65: 2011. https://doi.org/10.1063/1.342791

Stoy, R. D. 1989b. "Solution Procedure for the Laplace Solution in Bi-Spherical Coordinates for Two Spheres in a Uniform External Field: Perpendicular Orientation." *Journal of Applied Physics* 66: 5093. https://doi.org/10.1063/1.343737

Yan, B.-X., Y.-Y. Zhu, Y. Wei, and H. Pei. 2021. "Study on Surface Enhanced Raman Scattering of Au and Au@Al_2O_3 Spherical Dimers

Based on 3D Finite Element Method." *Scientific Reports* 11: 8391. https://doi.org/10.1038/s41598-021-87997-z

Yang, W.-H., G. C. Schatz, and R. P. Van Duyne. 1995. "Discrete Dipole Approximations for Calculating Extinction and Raman Intensities for Small Particles with Arbitrary Shape." *Journal of Chemical Physics* 103: 869. https://doi.org/10.1063/1.469787

CHAPTER 16

SUPER-RESOLUTION IN MICROSPHERES AND TRANSFER FUNCTIONS FOR EVANESCENT WAVES AND MICROSPHERES

16.1. Introduction

Any microscopic image can be magnified by using a microscope. However, observing sub-wavelength structures is difficult because of the *Abbe diffraction limit* (Lipson, Lipson, and Lipson 2010). According to this limit, light with wavelength λ, travelling in a medium with a refractive index n, and aperture angle θ, will form a spot with radius:

$$d = \frac{2(n\sin\theta)}{\lambda} . \tag{16.1}$$

The term $n\sin\theta$ is called the numerical aperture (NA), and the Abbe limit is on the order of $\lambda/2$. In the present chapter, I analyze the mechanism by which evanescent waves incident on a microsphere are converted into propagating waves. This conversion enables the microscope to produce high-resolution images, far beyond the Abbe limit.

16.2. Methods

The use of evanescent waves to increase the resolution beyond the Abbe limit can be related to the Helmholtz equation (Goodman 2017). In a homogeneous medium, this equation is given by:

$$(nk_0)^2 = k_x{}^2 + k_y{}^2 + k_z{}^2 \tag{16.2}$$

where $k_0 = 2\pi / \lambda_0$, λ_0 is the wavelength in vacuum, and k_x, k_y, k_z are the wavevector components. The evanescent waves that satisfy the relation

$$k_x{}^2 + k_y{}^2 > (nk_0)^2, \tag{16.3}$$

arrive at the microsphere with an imaginary k_z. An increase in the in-plane wavevector components $\vec{k}$ decreases the "effective" wavelength in the *x-y* plane, thus increasing the resolution.

As described in Figure 16.1, a dielectric microsphere with radius *R* and a refractive index n_2 is positioned above a thin object at contact point *O*. The medium between the object and the microsphere has a refractive index n_1. Parallel EM waves are transmitted in a direction perpendicular to the thin planar object, which may be transmitted as "propagation" and "evanescent" waves. The increase in resolution achieved by the microsphere is attributed to the conversion of evanescent waves into propagating waves.

We analyze the conversion of evanescent waves into propagating waves at a point *P* located on the microsphere surface. The angles of incidence and transmission for the EM waves entering the microsphere at point *P* are given as Θ_I and Θ_T, respectively. At this point, EM waves with wavevectors $k_x.k_y$ are arriving at the microsphere, where k_z is imaginary. As mentioned above, an increase in the in-plane wavevector component $\vec{k}$ reduced the "effective" wavelength in this plane, thus enhancing resolution. However, evanescent waves decay in the perpendicular *z*-direction. To "capture" the fine structural details carried by the evanescent waves, point *P* must be close to the contact point *O*, such that its perpendicular distance from the object is on the order of the wavelength. According to geometric optics, the microsphere has a spherical symmetry under rotation around the *z*-axis, which connects the center of the microsphere at point *C*, with the contact point *O*. The line connecting the center point *C* with point *P* forms an angle Θ_I with the symmetric *z*-axis.

As shown in Figure 16.1, the EM waves transmitted through the microsphere converge into a photonic jet (PJ), whose role in producing high resolution remains controversial. The original work demonstrating high resolution in microsphere imaging was made by Wang et al. (2011).

That analysis was based on a virtual image, as described by geometric optics. Since then, a large number of studies have been published on various effects in microsphere systems. In Figure 16.1, we describe a real image produced by a high-refractive index microsphere (e.g., Darafsheh et al. 2012; Lai et al. 2016).

The focusing of light in the microsphere system occurs within the nano-jet sub-diffraction region, which does not obey the classical laws of geometrical optics. This behavior might be explained by diffraction PJ effects (e.g., Lecler, Takakura, and Meyrueis 2005; Horiuchi 2012). Exact solutions for non-diffraction beams may be related to the central part of the photonic nano-jet (Durnin 1987; Ben-Aryeh 2017). Rigorous Mie theory predicts the interaction of light with spherical particles, and has been used to describe various properties of the PJs produced by microsphere systems (Lecler et al. 2019). The application of the Mie theory typically relies on numerical calculations, as it involves the summation of many terms and does not yield an analytical result. Using Mie theory, optical resonances in microsphere photonic nano-jets have been observed (Lee, Li, and Wang 2014). We analyzed the properties of evanescent waves produced by plane EM waves transmitted through a nano-corrugated metallic thin film, which carry information on the film's fine structures (Ben-Aryeh 2012). A microsphere located above the metallic surface collects the evanescent waves, which are then converted into propagating waves. The magnification of nanostructure images can be explained using optical principles, but the high resolution is attributed to the behavior of the evanescent waves. An approach for explaining the high resolutions obtained with microspheres was developed by using complex Snell's law (Ben-Aryeh 2016). Very high resolutions obtained with microspheres have also been reported in other studies (e.g., Yan et al. 2014; Wu and Hong 2021).

Maslov and Astratov (2024; 2019) studied the origin of super-resolution in microsphere-assisted imaging. Imaging based on the photonic nano-jet does not adequately explain the high resolutions obtained in microsphere systems. Similarly, the contribution of evanescent waves is insufficient to account for the observed resolutions, as their effect is relatively weak. Maslov and Astratov (2024; 2019) suggested a direct approach based on Maxwell's equations, incorporating optical principles such as point-spread functions in microspherical imaging.

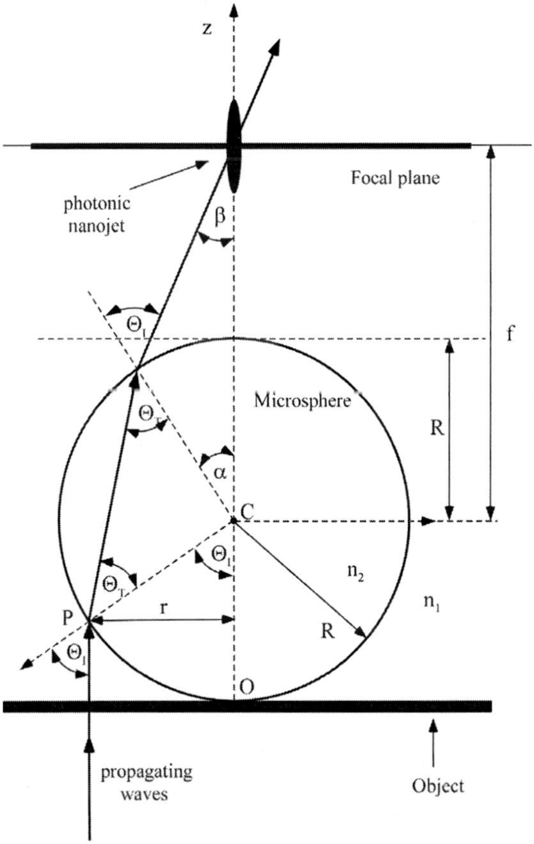

Figure 16.1: Transmission of EM waves through a Microsphere. Reproduced from Ben-Aryeh, Y. 2019. "Super Resolution of Nanomaterials and Quantum Effects Obtained by Microspheres." *Recent Progress in Materials* 1: 3. Licensed under CC.

Despite the limitations of existing models, there are strong arguments supporting the inclusion of evanescent waves in theories of super-resolution in microsphere imaging:

a) The propagating distance between the object and the microscope must be small enough to reveal the nanometric features. To generate super-resolution over large sample areas, the microsphere must be mounted on a frame that scans the sample in a step-by-step fashion (Huszka, Yang, and Gijs 2017).

b) The observed dependence of super-resolution on the radius of the microsphere and its index of refraction corresponds well with the properties of evanescent waves (Ben-Aryeh 2016).
c) Super-resolution is fundamentally related to information theory, where the fine structural details of the object are carried by the evanescent waves. Magnification and optical transforms alone cannot yield high resolution if the relevant information on the object's fine structures is not already present in the transmitted signal.

Propagating EM waves are transmitted through a thin object with a defined structure, whose image is produced by evanescent waves arriving at the microsphere surface, at point *P*, for example. The EM transmitted through the microsphere converge into a non-diffraction region known as PJ. I find, however, that there is an important mechanism which enhances the transformation of evanescent waves to propagating waves beyond the microsphere surface, one that lies beyond the scope of classical geometric optics and the application of complex Snell's law (Ben-Aryeh 2016). To explain this phenomenon, it will be helpful to compare the high resolution obtained by the microsphere with that obtained in other systems, as described as follows.

Experiments have shown that a thin film metallic grating with arrays of subwavelength holes can transmit light at certain frequencies with intensities an order of magnitude greater than the light incident on the area of these holes. Most investigators agree that these experiments, originally conducted by Ebessen et al. (Ebbesen et al. 1998; Ghaemi et al. 1998), are related to the coupling of light with surface plasmons. I have explained these phenomena as transmission enhancement resulting from the conversion of evanescent waves, entering the small holes, into propagating waves. This conversion occurs through the *convolution of the evanescent spatial modes with the plasmon spatial modes,* producing high spatial wavevectors and thus very small "effective" wavelengths (Ben-Aryeh 2008). Similar effects have been described by relating the 'tunelling' of evanescent waves into propagating waves to the convolution of the high spatial frequencies of the source with those of the detector (Ben-Aryeh 2006).

The super-resolution obtained by the microsphere is analogous to that in scanning near-field optical microscopy (SNOM), where the resonant effect at the tip detector enhances the super-resolution. In the dielectric microsphere, surface plasmons are not present, but other coupling mechanisms exist. In the present analysis, we follow the idea that the evanescent wavevectors are coupled to the microsphere through resonances, as described by Mie theory (Lee, Li, and Wang 2014). Mie theory is also used to provide a good description of the photonic nano-jet (Lecler et al. 2019). The enhanced conversion of evanescent waves into propagating waves, *after the transmission through the microsphere surface*, is obtained by convolution between the evanescent spatial modes and the transfer function of the microsphere modes (Goodman 2017). The use of such a convolution was already suggested in a previous article (Lee et al. 2013). The approach here, to microsphere-based high resolution, will be developed in the next section using two steps:

1) In the first step, we apply the Helmholtz equation by which large wavevectors are produced above the object (see Eq. (16.3)), leading to high resolution.
2) In the second step, after transmission through the microsphere surface, the conversion of evanescent waves into propagating waves is enhanced due to coupling between evanescent waves and resonances produced in the microsphere, for example, as described by Mie theory. In this stage, we do not rely on the Helmholtz equation. Instead, the coupling is described as a convolution between the spatial modes of the evanescent waves and those of the microsphere modes, represented by the *transfer function*.

16.3. An analysis of the super-resolution obtained in the microsphere system

As outlined above, we develop the analysis of the microsphere system into two parts:

a) First, we describe the propagation of evanescent waves, generated by fine structures on a thin planar object, to the microsphere surface. High image resolution at this stage arises from the behavior of large wavevectors governed by the Helmholtz equation.
b) Second, we consider the propagation of both evanescent and propagating waves, produced from the inner surface of the microsphere to the nano-jet. This propagation is described by the convolution of the EM fields with the microsphere modes, described by a transfer function (Goodman 2017). Although related

to Mie theory, this transfer function is more conveniently treated as an experimentally derived function.

16.4. The use of the Helmholtz equation for achieving high resolution by evanescent waves

Let us assume that the planar surface of the object lies in the plane $z = 0$, and that the EM field in this plane is given by its Fourier transform:

$$U(x,y)=\int_{-\infty}^{\infty}\int_{-\infty}^{\infty} u(k_x,k_y)\exp\left[-i(k_x x+k_y y)\right]dk_x dk_y \quad (16.4)$$

Then, the EM waves propagating from the planar surface of the object into a homogeneous medium, occupying the space $z>0$, with a refractive index n_1, are given by:

$$U(x,y,z>0)=\int_{-\infty}^{\infty}\int_{-\infty}^{\infty} u(k_x,k_y)\exp\left[-i(k_x x+k_y y+k_z z)\right]dk_x dk_y \quad (16.5)$$

Substituting Eq. (16.5) into the Helmholtz equation:

$$(\Delta^2+k^2)U(\vec{r})=0, \quad (16.6)$$

we get:

$$\left[k^2-(k_x^2+k_y^2+k_z^2)\right]U(x,y,z>0)=0 \quad (16.7)$$

Under the condition $k^2<k_x^2+k_y^2$, k_z becomes imaginary. In this case, we get the evanescent wave solution:

$$U(x,y,z>0)=U(x,y,z=0)\exp(-\gamma z);$$
$$\gamma=\sqrt{k_x^2+k_y^2-k^2}. \quad (16.8)$$

For evanescent waves, the field decays exponentially in the *z*-direction. The resolution obtained by the evanescent waves is limited by the lateral component of the wavelength, given by:

$$\lambda_T = \frac{2\pi}{k_T} = \frac{2\pi}{\sqrt{k_x^2 + k_y^2}} . \tag{16.9}$$

As the evanescent waves satisfy the equation $k^2 = k_x^2 + k_y^2 - \gamma^2$, Eq. (16.9) can be written as:

$$\lambda_T\left(evan.\right) = \frac{2\pi}{\sqrt{k^2 + \gamma^2}} . \tag{16.10}$$

The *minimal* value of λ_T for propagating waves is given by:

$$\lambda_{T,\min}\left(prop.\right) = \frac{2\pi}{k} , \tag{16.11}$$

since the minimum is obtained when $\vec{k}$ is in the *x-y* plane.

As the Abbe limit sets the resolution on the order of $\lambda / 2$, the increase in resolution achieved by using evanescent waves is given by:

$$F = \lambda_T\left(prop.\right) / \lambda_T\left(evan.\right) = \sqrt{\frac{k^2 + \gamma^2}{k^2}} = \sqrt{1 + \frac{\gamma^2}{n_1^2 k_0^2}} . \tag{16.12}$$

The distance of the point *P* from the planar object is denoted by $h = R(1 - \cos\theta_I)$ (see Figure 16.1). For a decay constant γ of the evanescent waves, represented in units of $\gamma / n_1 k_0$, the amplitude decay of the evanescent wave at point *P*, after traversing distance h, is given by:

$$\exp\left(-\gamma h\right) = \exp\left[-\frac{\gamma}{n_1 k_0} n_1 k_0 R(1 - \cos\theta_I\right]$$
$$= \exp\left[-\frac{\gamma}{n_1 k_0} 2\pi\left(1 - \cos\theta_I\right)\frac{R}{\lambda}\right]. \tag{16.13}$$

In our paper on complex Snell's law (Table 1, Ben-Aryeh 2016), we demonstrated the decay of the evanescent wave as a function of the parameters $\gamma/n_1 k_0$ and $(1 - \cos\theta_I)$, for a certain R/λ, representing the radius of the microsphere as equal to the product of the wavelength and a small number (Lecler et al. 2019). (Note that in Ben-Aryeh 2016, the notation α was used instead of θ_I). We found that the decay increases significantly with increasing θ_I, indicating that only a small distance around the contact point O effectively obtains high resolution through evanescent waves.

16.5. Microsphere imaging by a transfer function from the microsphere surface to the nano-jet

Let us assume that the EM field beyond the microsphere surface (comprising evanescent waves in a small region around the contact point O, and propagating waves over a larger region) is given by:

$$E'(x,y) = E_{\tan}(x,y)G(x,y), \tag{16.14}$$

where $E_{\tan}(x,y)$ is the tangential component of the EM field over the microsphere surface, and $G(x,y)$ is the 'transfer function' of the microsphere. This function can be expressed in terms of its Fourier transform (Goodman 2017) as:

$$G(k_x,k_y) = \int_{-\infty}^{\infty}\int_{-\infty}^{\infty} G(x,y)\exp\left[-i\left(k_x x + k_y y\right)\right]dxdy \tag{16.15}$$

The EM field $E_{\tan}(x,y)$ is preserved during the transmission through the surface (Ben-Aryeh 2016). This field includes both the evanescent and propagating wave components. It is important to note that $G(k_x,k_y)$ should include both evanescent waves for which $k_x{}^2 + k_y{}^2 > k^2$, and propagating waves for which $k_x{}^2 + k_y{}^2 < k^2$.

Due to the small distance between the point (x, y) on the microsphere surface and the contact point O, the evanescent field $E_{\tan,evan}(x,y)$ is approximately parallel to the x-y plane. This EM field is smaller than the evanescent EM field on the thin object by a factor approximately given by Eq. (16.13).

The Fourier amplitude $A(k_x,k_y)$ is described by the inverse Fourier transform of $E_{\tan}(x,y)$, which is given by:

$$A(k_x,k_y)=\int_{-\infty}^{\infty}\int_{-\infty}^{\infty}E_{\tan}(x,y)\exp\left[-i\left(k_x x+k_y y\right)\right]dxdy \qquad (16.16)$$

where k_x, and k_y are the wavevector components in the *x* and *y* directions. The convolution of the spatial wavevectors $G(k_x,k_y)$ and $A(k_x,k_y)$ is given by (Goodman 2017):

$$E'(k_x,k_y)=\int_{-\infty}^{\infty}\int_{-\infty}^{\infty}A\left(k'_x,k'_y\right)G\left(k_x-k'_x,k_y-k;_y\right)dk'_x\,dk'_y \qquad (16.17)$$

This convolution includes spatial wavevectors $k_x-k'_x$ and $k_y-k'_y$ that can be very large, corresponding to significantly reduced effective wavelengths. For example, if approximately, $k_x-k'_x=kn$ where *n* is a large integer, then the effective wavelength will be reduced to λ/n (see the analysis in Lee et al. 2013). We use the convolution effect described by Eq. (16.17) to obtain enhancement of the conversion of evanescent waves into propagating waves by reducing the effective wavelength. Although the modulation of the EM waves by evanescent waves is small relative to the total intensity, such modulation effectively produces high resolution. The spreading of the spatial modes by convolution is analogous to the point spread function described by Maslov and Astratov (2024; 2019). However, it is important to account for the super-resolution obtained at the object plane during the first stage of the imaging process, which is governed by the Helmholtz equation.

The above reduction of the effective wavelength by convolution was also described as a quantum effect, in which entanglement between *n* photons reduces the effective wavelength to λ/n (Ben-Aryeh 2003; Ben-Aryeh 2019). Such entanglement persists even after the transformation of evanescent waves into propagating waves.

16.6. Discussion and conclusions

The high resolution obtained in microsphere imaging is due to two factors:

1) The EM radiation transmitted through the object produces both evanescent waves, which carry information on the fine structures of the object (smaller than a wavelength), and propagating waves, which include the large image of the object (with dimensions larger than a wavelength). Since the evanescent waves decay over a distance on the order of the wavelength, only those close to the object effectively preserve fine structural information. However, by attaching the microsphere to a movable frame that scans the object's surface, it is possible to construct a high-resolution image over a large area.
2) By applying boundary conditions derived from complex Snell's law, we find that the perpendicular component of certain evanescent waves at the microsphere surface becomes real and thus propagating. While this effect is important, there is another important factor that should also be considered. The convolution between the EM waves, on the inner surface of the microsphere, and the transfer function enables propagation with small effective wavelengths (i.e., large wavevectors). This effect arises from the spreading of wavevectors introduced by the convolution. Because the transfer function also includes the wavevectors associated with evanescent waves, it enables these components to be transferred to the image without evanescent wave decay. Although Mie theory may be used to calculate the transfer function, such calculations are highly complex and usually give only numerical results. Therefore, we propose to use the transfer function as an experimental function, which may also account for the formation of the nano-jet.

Figure 16.1 gives only a geometric optics representation. The incident and transmitted angles, θ_I and θ_T are derived by Snell's law. Several features of the microsphere system can be interpreted using this geometric framework. For example, a microsphere with a low refractive index produces a diverging beam and a virtual image, while a high refractive index results in a converging beam and a real image, as shown in Figure 16.1. The angle β between the beam converging to the nano-jet and the symmetric axis, as well as the distance r from the point *P* to the axis, it can be obtained through straightforward geometric calculations. However, the super-resolution capabilities of the microsphere must be evaluated using the quantum analysis presented in this chapter.

References

Ben-Aryeh, Y. 2003. "Nonclassical High Resolution Optical Effects Produced by Evanescent Waves." *Journal of Optics B: Quantum and Semiclassical Optics* 5: S553–S556. https://doi.org/10.1088/1464-4266/5/6/002.

Ben-Aryeh, Y. 2006. "Tunneling of Evanescent Waves into Propagating Waves." *Applied Physics B* 84: 121–124. https://doi.org/10.1007/s00340-006-2220-3.

Ben-Aryeh, Y. 2008. "Transmission Enhancement by Conversion of Evanescent Waves into Propagating Waves." *Applied Physics B* 91: 157–165. https://doi.org/10.1007/s00340-008-2945-2.

Ben-Aryeh, Y. 2012. "Super-Resolution Observed from Evanescent Waves Transmitted through Nano-Corrugated Metallic Films." *Applied Physics B* 109: 165–170. https://doi.org/10.1007/s00340-012-5193-4.

Ben-Aryeh, Y. 2016. "Increase of Resolution by Use of Microspheres Related to Complex Snell's Law." *Journal of the Optical Society of America A* 33: 2284–2288. https://doi.org/10.1364/JOSAA.33.002284.

Ben-Aryeh, Y. 2017. "Nano-Jet Related to Bessel Beams and to Super-Resolution in Microsphere Optical Experiments." *EPJ Techniques and Instrumentation* 4: 3. https://doi.org/10.1140/epjti/s40485-017-0038-5.

Ben-Aryeh, Y. 2019. "Super Resolution of Nanomaterials and Quantum Effects Obtained by Microspheres." *Recent Progress in Materials* 1: 3. https://doi.org/10.21926/rpm.1903003.

Darafsheh, A., G. F. Walsh, L. D. Negro, and V. N. Astratov. 2012. "Optical Super-Resolution by High-Index Liquid-Immersed Microspheres." *Applied Physics Letters* 101: 141128. https://doi.org/10.1063/1.4757600.

Durnin, J. 1987. "Exact Solutions for Nondiffracting Beams. I. The Scalar Theory." *Journal of the Optical Society of America A* 4: 651–654. https://doi.org/10.1364/JOSAA.4.000651.

Ebbesen, T. W., H. J. Lezec, H. F. Ghaemi, T. Thio, and P. A. Wolff. 1998. "Extraordinary Optical Transmission through Sub-Wavelength Hole Arrays." *Nature* 391: 667. https://doi.org/10.1038/35570.

Ghaemi, H. F., T. Thio, D. E. Grupp, T. W. Ebbesen, and H. J. Lezec. 1998. "Surface Plasmons Enhance Optical Transmission through Subwavelength Holes." *Physical Review B* 58: 6779. https://doi.org/10.1103/PhysRevB.58.6779.

Goodman, J. W. 2017. *Introduction to Fourier Optics*. 4th ed. New York: W. H. Freeman.

Horiuchi, N. 2012. "Photonic Nanojets." *Nature Photonics* 6: 138–139. https://doi.org/10.1038/nphoton.2012.43.

Huszka, G., H. Yang, and M. A. M. Gijs. 2017. "Microsphere-Based Super-Resolution Scanning Optical Microscope." *Optics Express* 25: 15079. https://doi.org/10.1364/OE.25.015079.

Lai, H. S., F. Wang, Y. Li, et al. 2016. "Super-Resolution Real Imaging in Microsphere-Assisted Microscopy." *PLOS ONE* 11: e0165194. https://doi.org/10.1371/journal.pone.0165194.

Lecler, S., S. Perrin, A. Leong-Hoi, and P. Montgomery. 2019. "Photonic Jet Lens." *Scientific Reports* 9: 4725. https://doi.org/10.1038/s41598-019-41193-2.

Lecler, S., Y. Takakura, and P. Meyrueis. 2005. "Properties of a Three-Dimensional Photonic Jet." *Optics Letters* 30: 2641. https://doi.org/10.1364/OL.30.002641.

Lee, S., L. Li, and Z. Wang. 2014. "Optical Resonances in Microsphere Photonic Nanojets." *Journal of Optics* 16: 015704. https://doi.org/10.1088/2040-8978/16/1/015704.

Lee, S., L. Li, Y. Ben-Aryeh, et al. 2013. "Overcoming the Diffraction Limit Induced by Microsphere Optical Nanoscopy." *Journal of Optics* 15: 125710. https://doi.org/10.1088/2040-8978/15/12/125710.

Lipson, A., S. G. Lipson, and H. Lipson. 2010. *Optical Physics*. Cambridge: Cambridge University Press. https://doi.org/10.1017/CBO9780511763120.

Maslov, A. V., and V. N. Astratov. 2019. "Resolution and Reciprocity in Microspherical Nanoscopy: Point-Spread Function Versus Photonic Nanojets." *Physical Review Applied* 11: 064004. https://doi.org/10.1103/PhysRevApplied.11.064004.

Maslov, A. V., and V. N. Astratov. 2024. "Origin of the Super-Resolution of Microsphere-Assisted Imaging." *Applied Physics Letters* 124: 061105. https://doi.org/10.1063/5.0188450.

Wang, Z., W. Guo, L. Li, B. Luk'yanchuk, A. Khan, Z. Liu, Z. Chen, and M. Hong. 2011. "Optical Virtual Imaging at 50 nm Lateral Resolution with a White-Light Nanoscope." *Nature Communications* 2: 218. https://doi.org/10.1038/ncomms1211.

Wu, G., and M. Hong. 2021. "Optical Nano-Imaging via Microsphere Compound Lenses Working in Non-Contact Mode." *Optics Express* 29: 23073–23082. https://doi.org/10.1364/OE.426231.

Yan, Y., L. Li, C. Feng, W. Guo, S. Lee, and M. Hong. 2014. "Microsphere-Coupled Scanning Laser Confocal Nanoscope for Sub-

Diffraction-Limited Imaging at 25 nm Lateral Resolution in the Visible Spectrum." *ACS Nano* 8: 1809–1816. https://doi.org/10.1021/nn406201q.